Tommaso Trevisani

Land and Power in Khorezm

Farmers, Communities, and the State
in Uzbekistan's Decollectivisation

LIT

Cover Photo: A community elder who has been called in to give a blessing at a *pokaz* in Yangibozor. In the background the *hokim* (wearing the baseball cap) gives instructions to *fermer*s on the use of fertilisers on their land. (Photo: Tommaso Trevisani, 2004).

This work is a revised version of my dissertation manuscript, submitted to the Faculty of Political and Social Sciences at Freie Universität Berlin in 2008.

Zugl.: Berlin, Freie Univ., Diss., 2008

Gedruckt auf alterungsbeständigem Werkdruckpapier entsprechend
ANSI Z3948   DIN ISO 9706

Bibliographic information published by the Deutsche Nationalbibliothek
The Deutsche Nationalbibliothek lists this publication in the Deutsche Nationalbibliografie; detailed bibliographic data are available in the Internet at http://dnb.d-nb.de.

ISBN 978-3-643-90098-2

A catalogue record for this book is available from the British Library

©LIT VERLAG GmbH & Co. KG Wien,
Zweigniederlassung Zürich 2010
Klosbachstr. 107
CH-8032 Zürich
Tel. +41 (0) 44-251 75 05
Fax +41 (0) 44-251 75 06
e-Mail: zuerich@lit-verlag.ch
http://www.lit-verlag.ch

LIT VERLAG Dr. W. Hopf
Berlin 2010
Fresnostr. 2
D-48159 Münster
Tel. +49 (0) 2 51-620 320
Fax +49 (0) 2 51-922 60 99
e-Mail: lit@lit-verlag.de
http://www.lit-verlag.de

Distribution:
In Germany: LIT Verlag Fresnostr. 2, D-48159 Münster
Tel. +49 (0) 2 51-620 32 22, Fax +49 (0) 2 51-922 60 99, e-mail: vertrieb@lit-verlag.de
In Austria: Medienlogistik Pichler-ÖBZ, e-mail: mlo@medien-logistik.at

In the UK: Global Book Marketing, e-mail: mo@centralbooks.com

Land and Power in Khorezm
Farmers, Communities, and the State in Uzbekistan's Decollectivisation

From its Soviet beginnings, the Uzbek cotton sector was an integral part of the Communists' power system and a vehicle for the social and economic modernisation of the countryside. The newly independent state of Uzbekistan retained an undiminished interest in cotton but was confronted with a deep crisis and de-modernisation in agriculture. The state wanted to reform production without loosening its control over rural communities and revenues, so its agropolicies, while appearing to maintain continuity with the socialist system, did not follow the classic neo-liberal trajectory. Nevertheless, the agrarian structure and the social relations of agricultural production in Uzbekistan have been substantially reshaped. The state still sets production targets and keeps a firm hold over land ownership and regulation, but it has privatised the risks of production. Managerial responsibilities have been transferred to family-based production units, which has triggered a re-patriachalisation of family structures as the extended family, in a pattern reminiscent of pre-revolutionary labour relations, has become the implicit prerequisite for successful farming.

During the 1990s, collective enterprises in Uzbekistan were at first only superficially reformed into shareholdings designed to extinguish the debts of the kolkhozes. In the second decade of independence, substantial changes came with the dismantling of these collective enterprises and the establishment of commercial farms. In 2003, Yangibozor district, in the Khorezm region, became a trail-blazer for a decollectivisation policy that was implemented nation-wide in the following years. Land increasingly became an unofficial commodity in Yangibozor, and the district authorities profited from it most, thanks to their discretionary power to confer entitlements to state land. Opportunities to acquire land as individual farmers were greater for the former kolkhoz elites than for the former brigade farm workers, who were granted a new status as farm labourers *cum* subsistence-oriented peasants, with minimal access to agricultural land. From the dissolution of the collective farms emerged two lines of conflict in rural communities. One is between farmers and peasants, who can be seen as the winners and losers, respectively, in the agricultural reforms. The other unfolds among the so-called winners themselves – the new agricultural elites – who now must divide their loyalties between the political interest of the state production plan and their commercial interest in their own farms.

Power over land and resources, once a privilege of the kolkhoz chairmen, has moved up the command hierarchy to become the prerogative of the heads of the districts in Khorezm region. With decollectivisation, the inequality of a power system based on privileges and resource distribution that characterised the formally egalitarian Soviet kolkhoz has become even more pronounced. Reversing a pattern established during the late socialist period, when redistributive corruption served to justify local elites in the eyes of their communities, corruption has now become more individual, so that the new elites lack the legitimacy of their predecessors. Resistance, however, has emerged not among the reform 'losers' but among individual, entrepreneurial counter-elites whom district authorities attempt to control.

 **Halle Studies in the Anthropology of Eurasia**

General Editors:

Chris Hann, Thomas Hauschild, Richard Rottenburg, Burkhard Schnepel

Volume 23

LIT

Contents

	List of Illustrations	ix
	Foreword by Chris Hann	xi
	Acknowledgements	xiii
	Note on Transliteration	xvii
1	**Introduction**	**1**
	The Postsocialist Agrarian Question in Khorezm	2
	Reforms, Cotton, and Conflicts	6
	Approaching State, Society, and Politics	11
	In the Cotton Fields and Beyond	15
	Yangibozor	23
	Fieldwork	26
2	**Khorezm Region**	**31**
	Khorezmian Uzbeks	34
	Khorezmian Communities	40
	The Khorezmian Family, Past and Present	43
	Gender Relations	47
	Land Tenure and Rural Families during the 'Feudal Period'	48
	From *Yakka Joy* to *Posolka*	52
	Socialist Modernisation in Khorezm	57
3	**The Late Kolkhoz Years in Soviet Uzbekistan**	**65**
	The Kolkhoz and Sovkhoz in Late Socialist Khorezm	67
	A Local Perspective on Late Socialist Modernisation	74
	The Collective and the *Rais*	83
	The Cotton Scandal as Seen from Yangibozor	89
	The Legacy of Late Socialism	94

4	**Postsocialist Agriculture**	**97**
	Postsocialist Agropolicy and Land Reform	98
	New Regulations and Land Use Patterns for *Dehqon*s and *Fermer*s	106
	Land Use Practices for Freely Marketable Crops	110
	Privatisation	112
	Decollectivisation in Yangibozor	118
	'Fairness' and New Rules of Agriculture	121
	The Reshaped Command Hierarchy	123
5	**Decollectivisation, Labour Relations, and Kinship**	**133**
	New Cleavages in the Former Kolkhoz	134
	The Worsening of *Dehqon*s' Livelihoods	139
	Changing Labour Relations	142
	*Fermer*s, District Authorities, and Struggles over Crop Growing	153
	*Fermer*s: Risk, Opportunities, and Family Ties	157
	Decollectivisation and Kinship	163
6	**Rural Elites in Competition**	**165**
	Centre-Periphery Dynamics in Uzbekistan's Rural Sector	166
	The Battle for Cotton Transposed to Yangibozor	169
	Reproducing Dependency Patterns	171
	Old Established versus Newly Emerging Rural Elites	175
	Changing Patterns of Patronage	179
	Local Political Entrepreneurs	181
	From Cotton Scandal to Battle for Cotton	184
	Concluding Remarks	190
7	**Farmers, Communities, and the State**	**193**
	A Broken Social Contract?	195
	*Dehqon*s' Complaints	198
	*Fermer*s' Ambitions	203
	The District Authorities' Vision	207
	*Fermer*s, State, and Community	210
	Rural Society after the Kolkhoz	215

8	**Conclusion**	**219**

Appendix. Data on Cotton Harvests and Areas Planted in Cotton on Yangibozor Collective Farms in the 1960s, 1970s, and 1980s (tables 5-10) 227

Bibliography 233

Index 253

List of Illustrations

Charts

1	Land allocation in Yangibozor district, 2006	26
2	Evolution of land reform in Yangibozor district, 1980s to 2003	104
3	Organisation of farm production in Khorezm as of 2004	121

Maps

1	Districts of Khorezm region in Uzbekistan	32
2	Qalandardo'rman *sel'soviet*, 1932	59
3	Bo'ston *shirkat*, 2003	60
4	Collective farms of Yangibozor district	79
5	Land worked by brigades in Xalqobod *shirkat*	135
6	*Fermer* enterprises in Xalqobod MTP	136

Plates

1	Carrying out a household survey in Yangibozor district	19
2	Land measurers of Yangibozor district compiling maps	29
3	Wrestling match held during Navro'z celebration	37
4	Example of a traditional Khorezmian house	53
5	Old house in O'yrat village	54
6	Cotton gin at the processing facility in Yangibozor district	70
7	Rural landscape in Khorezm	100
8	Seminar in which district authorities give instructions to *fermer*s	128
9	List of *fermer*s closed down for legal infringements	131
10	Peasant with wheat harvest from his *tomorqa* plot	141
11	*Fermer* weighing cotton during the harvest	147
12	School boys resting after picking cotton for a *fermer*	150
13	Entrance to the former kolkhoz 'O'zbekiston'	173
14	Wall of propaganda	199
15	Village council document assigning community corvée to *fermer*s	213
16	Celebration of the fulfillment of the cotton quota at Bo'zqal'a MTP	215

(all photographs were taken by the author [2003-2004] unless noted otherwise)

Tables

1	Farm specialisation in Yangibozor district, 2004	119
2	Evolution of the command hierarchy in agriculture in Khorezm	125
3	Changes in the local political game in Yangibozor district	180
4	Relations of power and production in Khorezm	222
5-10	See Appendix	227

Foreword

Although Uzbekistan has by far the largest population among the Central Asian states that became independent following the break-up of the Soviet Union, the country remains poorly documented in Western anthropological literature. The historical importance of cities such as Bukhara and Samarkand and the Ferghana Valley's reputation for zealotry have inspired several studies of Islam after socialism, but all fieldwork-based research in Uzbekistan has been hindered by repressive conditions. Those who succeed in establishing close relations with people in Uzbek communities must take great care in publishing their results in order not to jeopardize the welfare of those with whom they worked.

Against this background, Tommaso Trevisani has produced the first detailed account of the transformation of agriculture in Uzbekistan, the sector on which most citizens depend for their livelihood. In the cotton-producing communities of the western province of Khorezm, the dismantling of Soviet institutions has been protracted. It began with the introduction of household-based sharecropping in the 1980s, continued with the replacement of the kolkhozes by *shirkat*s in the 1990s, and culminated only after the turn of the millennium with the emergence of the private farmer (*fermer*), who is still not the legal owner of the fields he cultivates. Uzbekistan's nominal embrace of a market economy has turned out to be highly selective. It has certainly not improved the position of the majority, whose labour is now exploited by new elites.

Cotton production in modern Central Asia has long been controversial. For some, it epitomizes colonial dependence, whether the centre was Moscow or Beijing. For others, investments in this sector provide evidence of a successful socialist development policy, the very opposite of an exploitative centre-periphery relationship. Whichever view one takes, it is undisputed that in Soviet times, Uzbek officials routinely exaggerated production figures in order to secure more resources from central planners. Trevisani argues that corruption on the part of local and regional elites also served the interests of ordinary collective farm members. The mysteries of what we might term 'socialist trickle-down' have now been replaced by the transparencies of a cutthroat individualism. The fermer, whose initial access to land usually derives from his political connections, is poised precariously between the state, his community, and an emerging stratum of absentee farm 'sponsors'. The power of this last group symbolizes the triumph of the town over the countryside.

Both inside and outside the domestic domain, the patriarchal extended family has been given a new lease of life. The dispossessed 'peasant' labourers are unable to mount any effective resistance to those who exploit them, but fermers, too, are vulnerable. Growing cotton according to an agreed 'business plan' allows them to eliminate risk, at least concerning input supplies, but it is much less profitable than growing rice. And land can no longer be 'hidden' as it was under the lax regimes of the kolkhozes and shirkats. The new farmers are thus obliged to become 'political entrepreneurs'. In addition to being clients of officials and urban sponsors, some continue to hold bureaucratic positions that, however poorly paid, enable them to maintain their vital networks.

Towards the end of his study Trevisani engages explicitly with two influential academic commentaries on the recent history of Uzbekistan. In the light of his evidence, Olivier Roy's theories concerning kolkhoz survivals as neo-tribal solidarity groups or an embryonic form of civil society have clearly lost whatever credibility they might have had for the 1990s. Deniz Kandiyoti's analysis of a 'broken social contract' is also criticized, because the conflicts of the new social structure are more complex than she envisaged. Although the state can on occasion crudely instrumentalise a 'communitarian ethos', not even the most powerful of the new farmers can ride roughshod over the norms of an older moral economy. One reason why the weapons of the weak remain so blunt in Khorezm seems to be that these new elites still honour their obligations to support local public institutions; they do not infringe on people's household plots, which continue to guarantee a minimal subsistence to all.

Trevisani's work is a valuable contribution to our knowledge of Uzbekistan, to the comparative literature on the 'postsocialist agrarian question', and to political anthropology in general. Despite the difficult conditions the author faced in the field, he manages to convey a keen sense of how different social groups are constrained by the new structures. The softer forms of Soviet corruption have been replaced by brazen clientelism, but there is no evidence that the present arrangements are more conducive to economic efficiency. For the time being, the transparent polarities of the new class relations are tempered by repatriarchalisation and the persistence of some vestiges of the old moral economy. One wonders whether this combination can possibly be a recipe for long-term stability.

Chris Hann

Halle, June 2010

Acknowledgements

This book would not have been possible without the help and encouragement of many people and the generous support of several institutions. The project started in autumn 2002 when Andreas Wimmer called me at the Zentrum für Entwicklungsforschung (ZEF, the Centre for Development Research) in Bonn to ask me to start a dissertation project under the supervision of Georg Elwert, who had been my teacher at the Institute of Social Anthropology at Freie Universität Berlin. Elwert's premature death in 2005, shortly after I completed my fieldwork in Khorezm, sadly interrupted a long, rewarding supervisory dialogue. I am grateful to Thomas Zitelmann for agreeing to take over the supervision of a dissertation project by then already well advanced. His advice and comments were invaluable as the book ripened into its present form.

I am also indebted to Chris Hann, both in his role as co-examiner, in which he offered constructive remarks on the thesis manuscript, and as the editor of this book series. I appreciate his accepting my manuscript for publication even though I was not affiliated with the Max Planck Institute for Social Anthropology in Halle, and for the institute's coverage of the copy-editing and publication expenses. I want to express my gratitude also to Peter Mollinga, who tutored me while I was a doctoral student at ZEF and who visited me in Khorezm during my fieldwork. I thank him for his interest in my work, our many discussions, and his offerings of advice, large and small, that found their way into this book.

The generous support of the ZEF/UNESCO Khorezm project, funded by the German Ministry for Education and Research (BMBF project 0339970A), provided me with a research context and infrastructure during my time in both Bonn and Khorezm. I am grateful to John Lamers, Cristopher Martius, Anja Schoeller Schletter, and Liliana Sin for their support and backing. I feel particularly indebted to Ro'zimboy Eshchanov, from Urganch State University, for his invaluable help in finding solutions to problems and facilitating my stay in Khorezm. Among my project colleagues in Khorezm, I wish particularly to thank Kirsten Kienzler for her friendship, generosity, and cooperation in the field. In addition, my fieldwork owed much to my colleagues Akmal Akramkhanov, Tamara Begdullaeva, Cristopher Conrad, Nodir Djanibekov, Oybek Egamberdiev, Irina Forkutsa, Mirzahayot Ibrakhimov, Asia Khamzina, Marc Müller, Gavhar Paluasheva, Aleksander Tupitsa, Gert Jan Veldwisch, Caleb Wall, Kai Wegerich, and Darya Zavgorodnaya Hirsch, and to many students at Urganch State University. I thank Zulmira Jabbarova, Rano Sabirova, and Elena Kan for their assistance during fieldwork and Natalia Shermetova for assistance in Tashkent. I thank Artur 'aka' for his tactful guidance and for his endurance, as he and the other drivers employed by the Khorezm project sometimes had to cope with long, burdensome workdays.

I cannot list everyone who, during my fieldwork in Khorezm, let me enter into their lives, shared their thoughts with me, or offered me their extraordinary hospitality. A special acknowledgement, however, must go to Ismoil Kutimov and Matyoqub Sherjonov. Their selfless support made a difference for my fieldwork in Yangibozor and for this book. I am also indebted to the staff members of the Khorezm Region Department of Agriculture, the Yangibozor District Department of Agriculture, and, as will soon become obvious to readers, the Yangibozor District Fermer and Dehqon (Farmer and Peasant) Association for the enormous amount of time they spent helping me. It is to 'my' *fermer*s in and around Yangibozor, though, that my largest debt of gratitude must go. Their names, or sometimes pseudonyms, appear in the pages to follow.

Any list of the scholars, colleagues, and friends who helped shape my thoughts for this book with their critiques, comments, advice, encouragement, or practical assistance will necessarily be partial. Nevertheless, I wish to express my thanks to Marco Buttino, Philipp Dorstewitz, Irit Eguavoen, Matteo Fumagalli, John Heathershaw, Irene Hilgers, Alisher Ilkhamov, Deniz Kandiyoti, Krisztina Kehl-Bodrogi, Kerstin Klenke, Alexander Morrison, Beatrice Penati, Niccolò Pianciola, Johan Rasanayagam, Philipp Reichmuth, Arnaud Ruffier, Armando Salvatore, Paolo Sartori, Anne Schober, Max Spoor, Julien Thorez, Massimo Toscani, Rano Turaeva, Piera Viale, and Lale Yalçın-Heckmann.

I did most of the final writing up of my dissertation in late 2007 and early 2008 in the southern Italian region of Apulia while working as a tutor for the University of Lecce. In Francavilla Fontana, where I lived, Andrea del Genio and the kind signora Margherita never stopped encouraging me through difficult times.

After submitting my doctoral thesis, I was granted a postdoctoral fellowship by the Gerda Henkel Foundation (project AZ 07/OP/07), which enabled me to revise the dissertation into this book while I was affiliated with the Stiftung Wissenschaft und Politik in Berlin (SWP, the German Institute for International and Security Affairs). I thank Volker Perthes and my colleagues from the research group Russia/Community of Independent States for enabling me to revise the manuscript for publication during my time at SWP.

I am grateful to Ingeborg Baldauf for her indispensable advice on the spelling and transcription of Uzbek words and for her opinions on a number of specific queries. Any mistakes are mine, as were the decisions taken.

I thank Jane Kepp for her meticulous copy-editing and Berit Westwood for her assistance in the final steps of manuscript preparation. Stephan Gourov helped me in reformatting and polishing all the maps and illustrations in the book.

I would like to thank my parents for their unceasing support over the years of my research project and for visiting me in Yangibozor. Finally, I thank Perla, my wife, for her endless support and commitment, for her comments, attention, discussions, and patience during the many years spent on this book, and for having proofread every sentence of it.

Parts of the book were published previously in different forms. I took some parts of chapter 4, including charts and tables, from a chapter I published in a book edited by Max Spoor ('The Reshaping of Inequality in Uzbekistan: Reforms, Land and Rural Incomes', in *Land Reform in Transition Economies: Contested Land in the 'East'*, pp. 123–137, London: Routledge, 2009). A lengthy ethnographic example used in chapter 5 was published in a volume edited by Deniz Kandiyoti ('The Emerging Actor of Decollectivisation in Uzbekistan: Private Farming between Newly Defined Political Constraints and Opportunities', in *The Cotton Sector in Central Asia: Economic Policy and Development Challenges,* pp. 151–174, London: School of Oriental and African Studies, 2007). A substantial part of chapter 6 was published as 'After the Kolkhoz: Rural Elites in Competition' in *Central Asian Survey* 26, no. 1 (2007): 85–104. A very early draft of chapter 7 was published as 'Rural Communities in Transformation: Fermers, Dehqons and the State in Khorezm', in *Patterns of Transformation in and around Uzbekistan,* eds. P. Sartori and T. Trevisani, pp. 185–215, Reggio Emilia: Diabasis, 2007.

Note on Transliteration

Having used different text sources, languages, and dialects, I opted for a compromise in this book between readability and formal correctness in spelling. For Uzbek words I use the official Latin orthography except when quoting other writers. I also spell recurring geographical and personal names in their anglicised forms – for example, Uzbekistan instead of O'zbekiston, Khorezm instead of Xorazm, and Khiva instead of Xiva. I use Uzbek forms for many recurring words taken from agricultural terminology, such as *fermer, dehqon, shirkat, ijara,* and *rais*, but for the sake of readability I use anglicised plurals for most Uzbek words. *Fermer* in the plural thus becomes *fermer*s instead of *fermerlar,* and so on. An exception is the plural of *rais,* for which, to avoid confusion with the English word 'raises', I use the Uzbek plural *raislar.*

For local toponyms, dialect words, and words taken from oral records or communication, I adopted the forms given in the Uzbek dictionary (*O'zbek Tilining Izoxli Lug'ati*) or in the Uzbek Soviet encyclopaedia (*O'zbek Sovyet Entsiklopediyasi*), whenever such forms were available. Otherwise, I opted for a transliteration that came phonetically close to the idiom (for instance *yoshulli, yumurtabäräk*). Russian words and names follow the transliteration system of the US Library of Congress, except when a widely accepted standard English spelling exists.

Non-English words are generally italicised only on first usage in a chapter, because there are so many of them and I use them frequently throughout the book. Exceptions are words that are spelled more or less identically to English words (for instance *rayon, tender, to'y*). These, together with words that appear only occasionally, are italicised in all instances.

Chapter 1
Introduction

Approximately 70 years after the forced introduction of collectivisation, decollectivisation marked the end of an era that deeply shaped rural Uzbekistan. In Central Asia, collectivisation was intended as both a measure to enhance agricultural production and a form of social modernisation of the peasantry. Decollectivisation, with its wide-ranging repercussions for rural communities, introduced changes that have reached beyond mere reform of the agricultural sector.

In this book I address the transformation of the agrarian structure of Uzbekistan by looking at the reshaping of livelihoods and agricultural practices in a cotton-growing district in the region of Khorezm. I assess continuity and change in social organisation, agricultural production, and the structure of power relations. Empirically, I focus on Uzbekistan's post-Soviet rural reforms – specifically, on the dismantling of the collective farms introduced under the Soviet Union. I also examine the new dynamics triggered by social transformations in rural society, which I contextualise within Khorezm's particular path of socialist and postsocialist modernisation.

Uzbekistan has followed a unique path of agrarian transition, which demands a specific investigation of social, political, and economic processes. The Uzbek case, as I show in the following chapters, confers on the term *decollectivisation* a meaning different from those applicable in other postsocialist scenarios. In this respect, this work is intended as a grass-roots-level account of rural people's perceptions of and attitudes towards the end of collective agriculture and the risks and opportunities they have faced with decollectivisation. In the shift from collective to post-collective agriculture, new conflicts have arisen and evolved between rural communities, agricultural producers, and local representatives of state authority. These conflicts are rooted in the political and moral economy of cotton growing, and digging into the kolkhoz past helps put recent struggles into historical perspective.

I carried out my research in Khorezm, a small regional administrative entity of the Republic of Uzbekistan, heir to one of Central Asia's khanates and later reshaped by the Soviet policy of national delimitation into today's administrative boundaries. In a broader sense, my research is about more than just Khorezm, for the problems and phenomena addressed tell something about ongoing dynamics reaching far beyond this small territory. The specificities of the cases I discuss may apply within certain cultural and geographical boundaries, but the national legal and political context is sufficiently uniform that my case study reflects a situation characteristic of all of rural Uzbekistan.

The Postsocialist Agrarian Question in Khorezm

My framing of the Uzbek (or Khorezmian) 'agrarian question' is an outgrowth of research on that theme by the Property Relations Group at the Max Planck Institute for Social Anthropology, as set out in the book *The Postsocialist Agrarian Question* (Hann and the Property Relations Group 2003). Originating in debates carried out at the turn of the nineteenth century, the agrarian question (Kautsky 1899) was first raised as an attempt to enlarge the field of Marxist analysis to include the previously neglected subject of the modernising countryside and the prospects of the peasantry and agriculture. The debate reflected the Marxist interest in forms of exploitation and revolved essentially around whether the purpose of modernisation was to convert 'feudal' rural society into a progressive and more productive class society, opposing capitalist landlords and a landless proletariat, or whether it was instead desirable that a more egalitarian (and socially conservative) order based on family farming should prevail in agriculture.

The Postsocialist Agrarian Question revisited this earlier debate in the aftermath of the socialist state experience in a variety of cultures and nation-states. It was concerned, 'on the one hand, with the production of agricultural goods and the efficiency of technologies, property rules, and markets, and, on the other, with the political, social, and moral consequences of these institutions in rural communities' (Hann 2003: 3). The authors stressed the great variety of forms and outcomes of postsocialist transitions, but they also reflected on resemblances and commonly recurring issues.

The prevailing image of the rural condition that emerged from this collection of studies and from other work was bleak.[1] Most ethnographers found themselves confronted with rural communities enduring economic devastation caused by the agricultural and industrial crises resulting from the end of socialist state systems. As the cost of living rose and social benefits

---

[1] Besides Hann and the Property Relations Group (2003), see Anderson and Pine (1995), Burawoy and Verdery (1999), and Hann (2002). On Uzbekistan, Zanca (2010) is among the works that most closely deal with this condition.

and services were severely curtailed, rural people's lives everywhere became more precarious.[2] Moulded on the precepts of the 'standard' (neo-liberal) prescription[3], postsocialist decollectivisation policies were aimed at disrupting the formerly collectively owned agricultural enterprises and redefining entitlements to landed property. In some cases these policies followed the ideal of making history 'reversible' (Giordano and Kostova 2002: 79) by re-establishing pre-existing ownership arrangements.[4] Few governments, however, could keep their implicit promises of welfare to their rural constituents. In most cases, decollectivisation resulted in benefits for the former local socialist agricultural establishment, members of which gained most from privatisation because they were able – thanks to their connections and skills – to convert their former political influence into economic capital (Lampland 2002; Swain 2003; Thelen 2003; Verdery 1998, 2003). To most other people, valued 'goods' such as land and assets turned into 'bads' (Verdery 2004: 156). That is, for many rural people, ownership of privatised land involved more risks and liabilities than opportunities for profit.

As Chris Hann (2005) pointed out, this process was controversial, and many anthropologists have emphasised the agency of rural communities in facing these undesired changes. Rural people experimented with new coping and survival strategies or asserted claims based on previously held collectivist values, protesting against government policies, as in Hungary (Hann 2004), or against the minority's profiting from the new reforms, as in Russia (Humphrey 2002). Decollectivisation engendered specific dynamics in different communities. Postsocialist rural societies became laboratories in which researchers could observe how new inequalities were being created, some old ones re-established, and some others abolished, and how rural communities reacted to these perceived changes.[5]

Many propositions and findings taken from the literature on postsocialism fitted my own experiences in Khorezm. Thelen (2003), for example,

---

[2] There were also cases in which rural communities faced violent conflicts and the situation obviously turned incomparably worse. In the literature on postsocialist conflict, affected rural communities have been studied particularly in the Caucasus and in former Yugoslavia. On the latter, see, for instance, Leutloff-Grandits (2006).

[3] The basic precepts of this position are that 'the means of production must now be privately owned, and market competition must prevail' (Hann 2003: 37; see also Hann 1998).

[4] This was not the case in Central Asia. Property restitution has been more topical in countries in which the time span between collectivisation and privatisation was shorter, as in eastern Europe. A wide range of anthropological literature deals with the peculiar problems arising from property restitution. To mention but a few sources, see Cartwright (2001) and Verdery (1996: 133–167) for Romania, Cellarius (2003) for Bulgaria, and Hann (2004) for Hungary, where one finds compensation instead of property restitution.

[5] On this subject, see also the programmatic paper written by Kressel (2004).

looked at 'winners' and 'losers' in postsocialist agriculture in Hungary and Romania, asking 'which new groups of producers have emerged on the basis of the postsocialist reforms', whether these were 'traditional family farms or totally new agricultural entrepreneurs', and 'how the excluded cope with and react against exclusion'. These questions proved highly relevant for Khorezm, and in adapting them to that setting, I was particularly interested in mechanisms of social inclusion and exclusion.

Yet Thelen's and other propositions in the literature on postsocialism were sometimes difficult to transpose to a context in which privatisation and reforms, as well as the basic organisation of society and the state, were far distant from the standard liberal blueprint of market and property. Although Hann (2003: 35ff.) understood postsocialism as a Eurasian paradigm, with ethnographic cases taken from as far east as China and Kamchatka, it seems that most generalisations rest on a core of empirical research conducted in eastern and central Europe in the 1990s. Situated at the very heart of Eurasia, Uzbekistan has been and continues to be in many ways different from those settings.[6] Identifying winners and losers in Uzbekistan's land reform requires a broadening of the notion of property, understood 'in terms of the distribution of social entitlements' to land (Hann 1998: 7). Control, use, and availability of land, more than ownership strictly speaking, are the sensitive variables that help explain why the ongoing reforms, despite their remoteness from the standard (liberal) model of market and property, have had such an effect on social relations.

One concept dealt with in the literature on postsocialist agrarian change that reveals a strong eastern and central European imprint is that of 'moral economy'. Drawing on the work of E. P. Thompson (1991), Hann (2003: 6) understood moral economy as the 'encapsulation of subjective experiences, norms, and values' that influence and interact with 'objective' economic transactions. More concretely, in postsocialist ethnographic narratives moral economy refers to the ways in which local communities relate to and react against changes in their economic institutions – changes induced externally, first by pre-socialist, then by socialist, and finally by liberal economic policies. This gives rise to conflicts, compromises, and forms of bargaining that unfold in the realms of legitimacy and ideology. They are expressed in the language of norms, values, 'ways of thinking', and religion.[7]

---

[6] As Kandiyoti (2002: 253–254) put it: 'Central Asian scholarship, always an outlier in the discursive analysis of state socialism, may diverge even further from Eastern European patterns as the post-Soviet trajectories of Central Asian societies unfold'.

[7] Harriss (2004: 161) calls for the recognition that rural development is a site of struggle, 'not least between ways of thinking'.

In socialist Europe, the moral economy found its key representative symbol in illicit but locally legitimated religious authorities who acted as custodians of the communal standpoint, especially in rural areas with strong religious traditions. Religion took on great importance in undermining state-directed policies and engendering intra-communal legitimacy struggles (see, for example, Abashin 2006; Gambold Miller and Heady 2003: 271). Much, too, has been written about the Muslim equivalent of this symbol in Central Asia, the unofficial imam or mullah, corresponding to the long-held image of Central Asia as a recalcitrant periphery, a bulwark of a traditional Muslim ethos that was potentially erosive or subversive of Soviet mores (see Bennigsen and Wimbush 1985). But as others have argued, this was a misleading image of local Muslim societies, a topos of Sovietology that failed to grasp essential characteristics of Soviet modernisation: 'The seven decades of Soviet rule left a deep imprint on Central Asia. The massive social engineering undertaken by the Soviet regime left very little unchanged, social classes were made and unmade, the terms of cultural debate were massively transformed, and the context in which Islam existed was radically altered' (Khalid 2007: 114).

Many recent anthropological monographs on Uzbekistan have highlighted the religious sphere as a space for contentions over legitimacy in communities (Kehl-Bodrogi 2008; Louw 2007; Rasanayagam 2010). My attempt to find what I called 'key holders of moral economy' in Khorezm revealed that local-level struggles over legitimacy were moving along different lines. I had expected to find that authoritative local spokesmen for the communities' interests referred, in their framing of the local agrarian question, to Muslim norms and values.[8] Instead I observed that rural communities facing decollectivisation seemed neither to rely on the language of Islam in their struggles for a 'just' agriculture nor to formulate close corporate interests.[9]

Visser (2008: 344) observed that in Russia, postsocialist farm restructuring was complicated by the fact that 'property and labour relations within the enterprises [were] interwoven with community relations based on being neighbours or relatives'. Therefore, despite neo-liberal reform precepts, boundaries between restructured kolkhozes and communities continued to be blurred. In Russia, postsocialist restructuring seems to have turned the kolkhoz into something more like a hacienda, in which the chairman 'pro-

---

[8] In this respect, my expectations were influenced by some literature about the Uzbek *mahalla* (Eckert 1996: 88ff.; Saktanber and Özataş-Baykal 2000), where religious spokespersons seemed to enjoy greater influence over community life than they did in Khorezm.

[9] Such interests would have been comparable, for instance, to those expressed by the *mir* (peasant commune) in postsocialist rural Russia (Gambold Miller and Heady 2003).

vides a patrimonial context and protection for the local community [in return] for their recognition of his right as owner and leader of the local community' (Nikulin 2003: 144). Reforms have failed to empower rural labourers as shareholders of formerly collectively owned goods (Visser 2008: 147), while behind a façade of rights granted only on paper, local state officials and farm directors have preserved pre-reform power relations and increased social disparity through 'Potemkin property rights' (Allina-Pisano 2008).

Despite differences in property relations and national economic policies, these symptoms by and large can be also found in Uzbekistan's path of agrarian transition. Expropriation and exploitation have occurred perhaps even more straightforwardly in Uzbekistan, less impeded by a nominal commitment to neo-liberal policy prescriptions. In Uzbekistan, however, the relationship between kolkhoz and community displays singularities rooted in the legacy of cotton. According to Zanca (2010: 149), in Uzbekistan 'the rais [kolkhoz chairman] literally embodies the historical antipathy between the power wielded by the state and the peasants' relative powerlessness', being one who 'lives among the people' while 'no longer being really of the people'. This relationship needs to be explored more fully, for with decollectivisation the established relationships between producers, communities, and authorities are increasingly challenged. In Khorezm, struggles over legitimacy were visible in changing relationships between officials still called *raislar* – the one-time kolkhoz chairmen and their postsocialist successors – and 'the people' (*xalq*) as decollectivisation disrupted patterns established with the late socialist kolkhoz.

Reforms, Cotton, and Conflicts

As in other countries of the former Soviet bloc, the scenario in Uzbekistan raises problems and perspectives that Wegren (1998: xiii) subsumed under his 'third wave' of land reforms, which were intended 'to destatize and to privatize land holdings in former communist countries'.[10] Such reforms had in common the economic rationale of 'enhancing agricultural production and performance by facilitating the rational use of rural labor and the efficient use of productive inputs', and they shared a political concern for the relationships 'between urban and rural dwellers, industrial and agricultural interests, the rural elite and the governing elite, and intra-agricultural inter-

---

[10] Wegren distinguished the third wave from a first wave of nationalisation and collectivisation of land, which took place in communist countries during the first half of the twentieth century, and a second wave of reforms backed by peasant movements' claims against large estates, especially in some Latin American countries.

ests' (Wegren 1998: xiii). Beyond their similar structural problems, rooted in their common legacy of socialist rule, postsocialist nations adopted reform trajectories with different priorities and contents, so that rural areas of the former socialist countries vary considerably today in the way they have been shaped by new land policies (Lerman, Csaki, and Feder 2004; Spoor and Visser 2001).

Yet two common traits appear to have emerged in all post-Soviet transition economies: increasing rural poverty and growing economic disparity. This is particularly true for Central Asian countries. Poverty and social disparity are rural phenomena, and they not only affect traditionally vulnerable groups such as the elderly and the children but also create a new class of 'working poor' (Spoor 2004: 47), among whom former kolkhoz workers represent, for different reasons, the 'worst off' (Spoor 2004: 62). In the Central Asian republics, land reform agendas gave priority to administrative and productive exigencies while neglecting social development objectives, which resulted in both social inequality and increased vulnerability, especially for the weak in rural society (Keyder and Kudat 2000). As a result, most peasants formerly attached to large, state-led enterprises (kolkhozes and sovkhozes) found themselves in what Griffin, Khan, and Ickowitz (2002: 20–26), drawing on Geertz (1963), defined as a process of 'agricultural involution'.

Uzbekistan is no exception to this trend, although relative to most other postsocialist countries, decollectivisation made its appearance there with a delay of a decade, and then only in a restrained form. Among the post-Soviet nations, Uzbekistan counts as one particularly resistant to reform (Gleason 2003; International Crisis Group 2004; Lerman 1998; Spoor 2003). Throughout the 1990s, Uzbek agricultural policies mandated only superficial reforms of pre-existing land tenure arrangements. Only in the second decade of independence did intensification of reforms lead to more substantial changes in the organisation of agriculture and in social relations.

In Uzbekistan, decollectivisation has not entailed privatisation of land. As in the past, land ownership has remained a prerogative of the state, which continues to intervene and to regulate agriculture, so that some of the most important features of the Soviet period have been maintained. The state-directed collective enterprises – the former kolkhozes, re-labelled *shirkat*s in the 1990s – were dismantled and their agricultural land distributed to newly established, smaller, family-based farmer (*fermer*) enterprises on the basis of long-term land leases. Peasants (*dehqon*s) still represent the overwhelming part of the rural workforce.

However different Uzbekistan's reform design may appear from decollectivisation in the European former socialist republics, the outcome has been similar, in that it has resulted in unequal access to land and unequal distribution of the risks and opportunities associated with the new conditions of land use. Despite the lack of liberalisation, decollectivisation has introduced new cleavages and conflicts into rural society.

Although Uzbekistan's agricultural sector has been far less reactive than those of its neighbours to the economic shocks that followed the demise of the USSR[11], it is often portrayed as having a particularly difficult rural condition, because it has a larger population and a higher rural population density than neighbouring countries.[12] According to Kudat (2000: 105), Uzbekistan's key rural development issues are its rapid population growth, shortages of water, natural gas, and credit, dilapidated social welfare system (such as education and social services), and problems related to the inadequate implementation of reform measures: land shortages, shortages in agricultural processing, and lack of participation in reform development and implementation. With Uzbekistan's rapidly growing rural population and prospect of shrinking natural resources (Micklin 2000)[13] – challenges inadequately tackled by the government, not least because of its 'unreceptivity to scientific-technical progress' (Trushin 1998: 288) – most literature has portrayed the country's rural development prospects as grim.

During the 1990s, the impoverishment of the rural masses (World Bank 1999), rising 'land hunger' (Kandiyoti 2003a), and growing social inequality, which fuelled a rural 'new poor' identity of dissent (Ilkhamov 2001), conferred on Uzbekistan's countryside the image of a problem area associated with instability and potential conflicts. This picture, coupled with the harsh circumstances that rural inhabitants faced during the post-Soviet transition, was aggravated by protests against the government on behalf of people the government calls 'Wahabis' – its radical Islamist opponents (see Khalid 2007; Rasanayagam 2010). In May 2005, the government's reaction of containment of opposition groups culminated in a massacre at Andijon when security forces fired into a crowd of protesters, killing hundreds (International Crisis Group 2005). Together these circumstances created a climate

---

[11] Kandiyoti (2003a: 226) argued that 'the initial decline in GDP experienced across the countries of the FSU was less steep in Uzbekistan due to the fact that the country produces a major export crop that could find alternative export markets'.

[12] Uzbekistan extends over 441,000 square kilometres. However, only 10 per cent of the land is agricultural (most of it must be irrigated), and 60 per cent of the population is concentrated in that area.

[13] On the basis of UNDP figures, Kudat et al. (1997: 109) wrote that 'if the current population growth rate of 3.4 percent continues, Uzbekistan's population can be expected to double in about the next twenty years'. At the time they wrote, Uzbekistan's population was given as 22.6 million people, 60 per cent of them in rural areas. Today estimates are about 27 million people.

of fear and perceived danger that has influenced the literature on rural Uzbek society. Whether implicitly or explicitly, many analysts see Uzbekistan's stability problem as rooted largely in the unresolved and growing agrarian question, because of which the authoritarian state is being challenged from the grassroots level.[14]

Influenced by this literature, my first idea of the post-Soviet Uzbek countryside fitted what Foster (1973: 35ff.) called the 'image of limited good'. That is, I conceptualised a static – in the case of Khorezm, even declining – system in which, after a Soviet 'era of plenty', available resources had become more limited and struggles over them more competitive. With Foster's paradigm in mind, I began my research with the assumption that new conflicts over resources were emerging in the local agricultural setting, with actors trying to outcompete one another for the use of scarce and precious goods such as land, water, access to markets, and inputs. The social situation I addressed awoke expectations of grassroots-level dissent, contestation, or possibly confrontation against the authoritarian state. Consequently, I expected to find patterns typical of the repertoire of peasant resistance theory (Scott 1985) – forms of 'peasant revenge' (Kitching 1998) or of 'socialist resistance' (Creed 1999: 230) – and I envisaged that a conflict framework would be highly relevant.

Ultimately, however, my interest in peasant resistance patterns proved to be overcharged with misleading expectations about rural people's grievances and readiness to protest. Upon a closer look, the social situation appeared less explosive than I had hypothesised from afar. Instead, as Elwert (2001) pointed out, the social organisation of risk perception seemed to converge on the avoidance of conflict, with the 'voice option' (Hirschman 1970) – the option of outright resistance against decollectivisation – bearing risks unmatched by its potential benefits. In the end I found that the social conflicts arising with decollectivisation in Khorezm were of a different kind and demanded a deeper understanding of local patterns. They contradicted the assumption of a profound alienation of the rural population from the political regime and confirmed the need to overcome simple models of totalitarianism in understanding the socialist and postsocialist Uzbek state (see Hann 1993, 2004).

Recent anthropological approaches to conflict acknowledge that conflicts follow socially ordered paths and are crucial for the creation and channelling of social cohesion (Eckert 2004; Elwert 2001). Bierschenk and

---

[14] This threat to stability was almost invoked by Lubin and Rubin (1999). On the nexus between the agrarian question and political stability, see also Fathi (2004). Western criticism of the Uzbek agricultural sector has especially highlighted the issue of child labour in the cotton sector (Kandiyoti 2009).

Oliver de Sardan (1997) emphasised the methodological usefulness of their approach to conflict. According to them (1997: 240), 'to speculate on the existence of consensus is a far less powerful and productive research hypothesis than to conjecture the existence of conflicts. Conflicts are the preferred indicators of the functioning of a local society. They are also indicators of social change and hence particularly important for the anthropology of development'. Their approach encompasses the recognition that rural development in general and development projects more specifically are arenas in which 'strategic groups' (Bierschenk and Elwert 1993; Evers and Schiel 1988), composed of actors who share a common interest in appropriating resources, exert influence, apply pressure, and struggle to meet their interests and those of their constituencies.

This underlying understanding of conflict and strategic groups seems to be shared by Harriss (2004: 161) when he writes that rural development can be seen as 'a political arena in which various actors negotiate and struggle, over and in the context of institutions (rules, norms, conventions) which regulate access to resources of different kinds, constrained variously by differences of power'.

In Uzbekistan's agricultural reforms, land and power turned out to be the key domains determining rural development. This statement is in line with observations made by Eckert and Elwert (2000) on the evolution of the Uzbek land tenure system. They pointed to disputes over land distribution 'triggered by occurrences of corruption' (2000: 37) and to 'potentially conflictive clientelist networks competing over scarce land' (2000: 47) as the prominent motives for conflict. Reminiscent of Boissevan's (1974) clientelist politics of network and coalition building, the Uzbek land reform scenario suggests the relevance of elite conflicts in local power relations. More generally, Eisenstadt and Roniger (1980) recognised in the 1970s that competition among elites for control over resources was a pattern usually generating patron-client relationships.[15] In their attempt to outline the basic characteristics and the worldwide variability of patron-client relations, they also mentioned the role of state bureaucracy and agriculture in the USSR (Eisenstadt and Roniger 1984: 157ff.), but they provided no specific reference to regional peculiarities and differences within the Soviet Union.

---

[15] These conditions 'generate, within each category of major social actors in these societies, but paradoxically enough especially among the higher groups and the incumbents of the center, a potential competition between members of each category with respect to their possible access to needed resources and to the positions that control these resources' (Eisenstadt and Roniger 1980: 70).

It was precisely in those years, however, that Soviet public opinion discovered the extreme relevance of patron-client networks in Central Asia, when the 'cotton scandal', or 'Uzbek affair', gained visibility:

> The 'Uzbek affair' of the late 1980s disclosed illegal inflation of cotton output reports in Uzbekistan, by as much as 4.5 million tons between 1978 and 1983. 'Insurance sowing' was discovered on around 200,000 ha. (almost 10 percent of the area of cotton sown in the republic). Some farm leaders, especially under Brezhnev's 'stability in cadres policy', moved beyond simple falsification of the plan and on to personal aggrandizement. The 'Uzbek affair' investigation revealed the endeavors of a number of petty, corrupt tyrants on kolkhozy and sovkhozy (Thurman 1999: 41).

The cotton scandal revealed the existence of a pervasive patronage system based on bribe-taking and the manipulation of production figures, showing that members of patronage networks colluded in defrauding the central government at all levels of the agricultural production system and the party hierarchy. In Soviet Central Asia, patronage networks served as bases from which to allocate scarce resources and political favours and to preserve regional political identity (see, for instance, Jones Luong 2002). Unfortunately, no studies have previously shed light on the nature of the constraints, obligations, and 'ethics' of these patronage relations with the accuracy of ethnographic work. This is all the more regrettable because this period of Central Asian history seems in many ways to have laid the foundations for contemporary relations between power and agriculture, centre and periphery, and rural and urban worlds and for the shaping of identity discourses straddling the Soviet and post-Soviet periods. Patronage patterns in Central Asia, although they are systematically assumed or taken for granted in a broad range of literature, have almost never been studied empirically. By revisiting the cotton scandal from a local perspective and from a moral economy approach, I attempt to develop a new frame for addressing the social and moral conflicts permeating Uzbek agrarian change today.

Approaching State, Society, and Politics

As acknowledged by Jones Luong (2004), Central Asia in many respects represents a case sui generis of state-society relations. There the Soviet state 'not only succeeded in profoundly transforming social and political organization ... but also in blurring the boundaries between state and society in distinctive ways' (Jones Luong 2004a: 4). Behind the Iron Curtain, in eastern and central European countries a civil society based on the model of voluntary association (Putnam 1993) could flourish and hold influence in political processes. In Central Asia the situation developed differently. There, even since the end of the Soviet Union, civil society has remained fragile and marginal, relegated to a subaltern condition of 'non-state structure parallel to the state' (Mandel 2002: 292), with state power opposing its full endorsement (Holt Ruffin and Waugh 1999).

In Uzbekistan the fracture between 'civil society' and 'political society' is particularly acute (Stevens 2007). The picture is not only different from that in most other postsocialist countries but also at variance with many of the lessons learned from other developing countries (Migdal 1988) and, despite obvious surface analogies, with the institutional landscapes and political cultures of neighbouring Middle Eastern Muslim societies, with which Central Asia is sometimes associated (Eickelman 2002).[16] Characterised by blurred boundaries between state and society, by weak participation of societal institutions in the public sphere, and by a submissive attitude on the part of the social domain towards the domain of the state, dynamics that elsewhere occur between state and society in Central Asia turn into struggles between different actors of the state, thus becoming struggles within the state (Jones Luong 2004b).

Today, a conceptual framework overwhelmingly occupied by the presence of the state has created fertile soil for culturalist explanations, which tend to portray Central Asian societies as 'deep-frozen' and immutable.[17] Besides oversimplifying the historical facts[18], the attractiveness of this and of other, more sophisticated but essentially related arguments rests on lack of knowledge about actual, ongoing processes of social change on the ground.[19]

Central Asia is relatively new to socio-anthropological research as it is understood in Western universities, which raises peculiar problems of orientation unknown to fieldworkers operating in areas with well-established research traditions. Although the Soviet school produced a massive amount of detailed information about Central Asia's traditional societies, this ethno-

---

[16] In terms of political culture, I think particularly of Bayat's (1997) 'street politics' and Tapper's (1990, 1997) historical perspective on 'tribe-state relations', both key contributions to the study of sociopolitical dynamics in Iran.

[17] An example is Starr's (1999) misleading argument ascribing the perceived weakness of (civil) society to a supposed collectivist tradition that discourages civic engagement, an argument that ultimately ends up justifying Wittfogel's (1957) oriental despotism argument.

[18] The problem with the oriental despotism approach is that it does not take political processes seriously enough, resulting in an over-schematic understanding of historical events. See, for instance, Barfield (1990) and Manz (1994) for an informed look at long-time political dynamics in Central Asian history. For a more recent period, see Yaroshevski (1984) for a description of the complexity of political struggles in the Khorezm oasis before the consolidation of Soviet power.

[19] In this regard I discuss the *querelle* between Kandiyoti and Roy in chapter 6.

graphic material resulted from a discipline that differed from Western social anthropology in its underlying models, methodological assumptions, and 'styles of thought' (Dragadze 1984). Today this complicates the interpretation of its achievements, and linking up with its scholarly tradition becomes problematic.[20] In Western academic discourse, 'criticism of the Soviet school of ethnography often becomes an exercise in setting up and beating a "straw man", whereby the basic tenets of the theoretical concepts developed and utilized within the Soviet tradition are oversimplified and even distorted' (Suyarkulova 2008: 398). Therefore, it is difficult today to make use of explanatory models and theories that were locally tested by earlier generations of anthropologists. Although since the beginning of the 1990s much anthropological research has been conducted in the area through extended fieldwork, Central Asia (and Uzbekistan) has maintained a marginal role within the discipline, in comparison with other, better-studied areas of the world, which have also become more influential for their contribution to general theory-building.

Following a classic quotation from Bailey (2001 [1969]: 1), one should 'think of politics as a competitive game. Games are orderly'. This perspective has seldom reached scholarship on Central Asian political systems. Although Sovietologists elevated Central Asia's political dynamics to one of their main research interests, the knowledge they produced about local political games of the Soviet era was rather shallow. Sovietological approaches had their own biases, their view was external and limited, and they failed to grasp many of the locally relevant processes.[21] Adopting opposed but in some sense complementary perspectives, Western Sovietologists and Soviet ethnographers impoverished social analysis and failed to appreciate the actual dynamics of change in Central Asia (Kandiyoti 2000: 53). Still, their analytical legacy has remained powerfully influential even since the demise of the Soviet Union. Today it can hamper attempts to think about ongoing social processes in more productive ways.

One legacy is the 'dual economy model' adopted in many analyses of socialism and postsocialism, which tends to artificially separate Soviet and post-Soviet economic reality into two disjoined realms, one of officialdom and the other of informality (along the dichotomies between formal and

---

[20] As put by Rasuly-Palaczek (2006: 10): 'Actual research results are comparatively few, research institutions and research traditions are not well established, and the rejection of Soviet-era findings leads many social anthropologists interested in Central Asia to feel that they have to start from scratch'.

[21] The emblematic case for this approach towards local power and ethnicity issues can be found in Rywkin's (1985) analysis of regional- and district-level party staffing in Uzbekistan, based merely on the changing names in the lists of party officials of given administrative bodies.

informal, official and shadow, and legal and illegal realms). By engaging with the modes in which households related to these supposedly distinct economic sectors in Uzbekistan, Rasanayagam (2002a: 68–69) showed that this separation in reality is inadequate, because 'at the level of local interactions these distinctions break down', and 'local modes of interaction' prevail. Rasanayagam (2002b, 2003) described how, from a fieldwork perspective on the everyday lives of rural households in Andijon, community, market, and state represented different interactional spheres with distinctive logics and attributes, a distinction that he showed to make more sense than the one between formal and informal institutions.

I had a similar experience in attempting to conceptualise state-society dynamics from within the district of Khorezm where I did most of my fieldwork. Instead of leading me to an abstract state-society dualism, my fieldwork evidence led me to a more pragmatic tripartition between rural communities, agricultural entrepreneurs, and local state officials, the social entities and actors that mattered in the local political game. They correlated, respectively, with a household-based, a commercial, and a state-ordered 'form' of production, with distinct economic features (Veldwisch and Spoor 2009). I unravel various aspects of the relationships among these three reciprocally interfacing elements of rural society by looking at their roles and interactions during decollectivisation.

Bellér-Hann and Hann (1999: 2) argued that 'sustained local-level investigations offer the best prospects for transcending the state-society dichotomy, through the identification of more specific groups within each and by showing how some people straddle the divide and bring it directly into question'. In my search for adequate theoretical frameworks, I found that their analysis of the situation of local state officials in rural areas of southern Xinjiang, China, mediating between the interests of peasant communities of their own ethnic background (Uyghur) and their Han-Chinese superiors, seemed better to suit the dynamics characteristic of my context than did many other central and eastern European postsocialist privatisation scenarios. The Uzbek SSR and China's autonomous Uyghur region were both peripheral regions situated at the margins of state-led socialist modernisation plans, relegated to the role of cotton producers by their 'foreign' (Russian, Chinese) political centres. Both were unequally compensated for their produce with subsidies and investments, and each in its own way gave rise to centre-periphery struggles with its socialist 'elder brother'. Although with Uzbekistan's independence from Moscow this macro-level parallel has come to an end, what are still similar are the micro-level coping strategies in a command economy framework based locally on mandatory cotton production schemes and state control, resulting in specific interaction patterns between state officials and agricultural producers. In southern Xinjiang, agricultural producers 'attempt to subvert controls and to win the connivance of local officials in strategies of evasion' (Bellér-Hann and Hann 1999: 3), similar to what happens today in Khorezm, even though Tashkent has taken the place of Moscow in the agricultural command hierarchy. It is precisely the post-independence transformation of these strategies of subversion, coping, and evasion, which characterise rural relations, that I focus on in this book.

Kandiyoti (2002) argued that a political economy framework for looking at the problems and survival strategies of primary producers offers a better approach to relevant processes in Central Asian states than do dependency or postcolonial theory approaches to state-society dynamics. This view was shared by Kressel (2004: 222): 'Both postmodernism and postcolonialism tackle problems within the history of anthropology that have no relevance to the study of communities of the former Soviet Union'. According to him, a political economy–minded 'neo-Marxist search for (material) power conflicts and Weberian attempts to spot culture differences that can explain observed economic gaps … in erstwhile communist societies' is a more promising way to reach an understanding of ongoing dynamics (Kressel 2004: 221). With new policies redefining previously established 'social contracts' between rulers and ruled, what is needed is a theory capable of addressing change in a less state-centric perspective. I follow Kandiyoti in recognising that at the interface between state, society, and politics, what the anthropology of Central Asia needs is approaches better able to grasp the discontinuities of the ongoing transformation than has been done so far.[22]

In the Cotton Fields and Beyond

Khorezm has long ceased to be a pre-industrial society. Soviet and post-Soviet societal developments have shaped it deeply and, as I show later, rendered it a 'complex society' (Banton 1966) of a particular kind. In approaching contemporary Khorezm, an in-depth study of the complexly articulated cotton sector proved to be a time-consuming but necessary precondition for any broader socio-political analysis. To be able to make more qualified inquiries into other aspects of rural life required an involvement with the economic, technical, legal, and organisational aspects of Khorezmian agriculture. Although these research objects might appear a bit

---

[22] 'Neither an understanding of pre-Soviet and Soviet institutions nor an analysis of changing legal frameworks can provide us with an adequate apprehension of these evolving realities. These can only be captured through detailed ethnographies that reveal both the intended and unintended effects of these policies, as well as the responses of those who are at the receiving end' (Kandiyoti 2002: 251).

dry and technical, it is illusory to address agrarian change anthropologically without first considering what most determines rural people's lives in Khorezm. My aim was to look at the cotton sector through the eyes of an anthropologist: how it is lived and perceived, and how to describe 'from within' what so far, with few exceptions, has been discussed mainly from a macro-level perspective, as viewed 'from outside' (Kandiyoti 2007: 7).

My research, however, was not intended as a mere bottom-up view of agrarian relations. Kressel (2004: 220) suggested that today 'a grasp of the regional and national dimensions of rural communities in ... [postsocialist] countries presupposes a genuine attempt at "studying up" or "sideways" in addition to "studying down"'. Accordingly, I looked at the making of new constellations of power in the local rural reform scenario by focusing on emerging actors, competing power holders, and peripheral counterclaims to centrally directed policies. Rather than emphasising decay and impoverishment in rural Uzbekistan (Zanca 1999, 2010), I was interested in its new vital impulses and dynamics. The most common reason so few attempts have been made to study postsocialist elites from an anthropological standpoint is the difficulty of accessing them.[23] This is especially valid for Uzbekistan, where the hermetic political situation reduces any room for 'upward' manoeuvre in fieldwork.

Affiliating myself with a German-Uzbek ZEF/UNESCO research and development cooperation project was a considered choice, given the crucial importance of local elites in Uzbekistan's rural reform and the methodological challenge posed by the difficulty of gaining access to them. Embedding my fieldwork in such a project was necessary for dealing with agrarian change in Uzbekistan, but it also influenced my research approach, in that I had to come to terms with the project's developmental and technical discourses and with a project environment that sometimes imposed restraints on my 'anthropological freedom'.

Because the government considered rural development its foremost prerogative, the project was dominated by 'technocratic discourses' (Bierschenk 1988: 153) and left no space for participatory or community empowering agendas. The project personnel had little interest in or awareness of what Bierschenk (1988) called 'arenas of negotiation for strategic groups', so misunderstandings between anthropologists and development operators were common (Olivier de Sardan 1995: 189ff.).

---

[23] 'Despite all good reasons for "studying up", why are these attempts so few in number? The most frequent reason is 'access': the rich and powerful are out of reach. They are too busy to give anthropologists their precious time. Furthermore, they do not want to be studied. In fact, it could actually be physically dangerous to explore the secret sources of their wealth' (Kressel 2004: 221).

The declared aim of the project was to contribute to an 'economic and ecological restructuring' of Khorezm (Vlek, Martius, and Lamers 2001). This meant that research by doctoral students was designed to probe techniques, develop technologies, and produce knowledge in the form of data that would be used, during a future implementation-oriented phase, to build a model of ways to modernise the local agricultural system. In essence, the project was conceived of as a technology transfer project, with the typical imperfections and short circuits that critics have described (see, for instance, Nolan 2002; Rottenburg 2002).[24]

The possibilities for integrating my results with those of the project's other research modules and for incorporating socio-anthropological data into its ultimate model were limited. The other research modules were far more uniform in their data production (laboratory, field experiment, quantitative-statistical data gathering), in their logistical needs (office facilities, field equipment, local enumerators, partners and supervisors from universities and research institutes), and in their utility for the project (raw material for model building). For me, unlike for my colleagues, connections with local scientists proved difficult to make. Other dissertation projects were easily linked with and embedded in the local setting, but my research was not, and it took considerable time for me to find both an entry point into the controversial, politically sensitive issue of land redistribution and legitimacy in the eyes of my interlocutors for my being in the field and asking questions.

In the end I decided as much as possible to 'play the project card' vis-à-vis my informants by consciously enacting the role of the foreign development expert seeking consultation, cooperation, and amity from local partners and authorities. Eventually I became a 'colleague' of some low-profile district administrators, and the staff of the local branch of the national Fermer and Dehqon (farmer and peasant) Association sometimes introduced me that way to other officials when we went together to do work in the offices of Khorezmian agricultural management. This legitimated my interactions with authorities and helped in accessing data, knowledge, and practices usually out of reach in village-centred anthropologies of Uzbekistan. However, I had no mandate from the project to operate in such way, nor did anyone have a clear notion of how I should operate in such a difficult fieldwork environment (Wall and Mollinga 2008).

In rural Uzbek society, the authoritarian nature of the government is manifested in a command hierarchy that, according to Ilkhamov (2000), 'dwells rather on the fear of administrative sanctions than on economic

---

[24] One dissertation written as part of the project was concerned with a critical study of the project as an instrument of development and as a system of knowledge management (Wall 2008).

incentives'. Threats of coercive measures by the *militsiya* (police, henceforth militia) and the *prokuror* (public prosecutor) represent the crudest levers of control over agricultural producers and can be used as a last resort to check non-compliance with the administrative hierarchy or even hostility against the government. On a more basic level, a variety of 'soft' means of control and intimidation influence people's behaviour in everyday life and keep them on the track designed by government officials. These circumstances generate a climate of fear and mistrust that has methodological repercussions for the anthropologist who attempts to conduct interviews about allegedly political topics such as reforms, rural incomes, and decision-making in agriculture.

It could be dangerous for people to express opinions that diverged from the official governmental position on certain issues, and the answers I received tended to be standardised according to the interviewee's place in the social hierarchy. Elusive or prescriptive answers to my questions were the rule in scheduled interviews held with local authorities, village officials, members of district and regional administrative bodies, farm managers, and other private and public personalities. This was also the case with ordinary citizens, except for a few whose trust I gained through repeated interviews, friendship, or their deeper sympathy with my scientific concerns.

Typically, officials in charge were inclined to portray work processes and reform outcomes in an embellished way, whereas dismissed or retired ones tended to speak more about problems and shortcomings. For most interlocutors, policy failures and controversies with authorities were taboo subjects. People's trust and confidence were necessary premises for research into land entitlements, agreements, and ownership patterns. External observation was of limited help and could not substitute for people's willingness to cooperate, because there was no land reform that could be seen, measured, or followed outside the words of the interlocutors and of mostly inaccessible documents.[25]

These circumstances made fieldwork methodologically challenging and often frustrating at the beginning. As often happens when anthropologists try to 'stick their noses' in local affairs, my initial field visits and encounters with public administrators met with evasive behaviour resembling a 'wall of indifference'. On paper, permissions to carry out research were released reluctantly; people were hesitant to collaborate because of the sensitivity of the topic. In interviews, answers were likely to follow the prescriptive positions of the government, even when reality manifestly

---

[25] Wall and Overton (2006: 65), who conducted research for the same project, commented: 'The disrespect for human rights in Uzbekistan, combined with the politicised nature of the rural economy ... made confidentiality vital in this research'.

contradicted them. Fearful officials shifted the responsibility for answering my questions to the few key actors holding power, who in turn had little interest in wasting their time on me. Even if my affiliation with the project gave me some credentials and a role that officials could not entirely ignore, I could offer little more than gratitude for their support. In light of this, the timely help I received from some people during my fieldwork was a great sign of generosity and commitment.

Plate 1. Carrying out a household survey in Yangibozor. Pictured are a village official (*elatkom*) who introduced me to the selected household, a militia officer overseeing my activity, and a resident who was curious about what I was doing. The car with the UN number plate was driven by a driver employed by the project.

In the hope of getting around these difficulties, I tried to access a village informally when I met a local state official who also ran his own farm and who declared himself willing to introduce me to local agriculture. By following him in his everyday affairs during my first two-month stay in the region in spring 2003, I learned a great deal about what I was interested in. In his perception, however, my interest in issues he judged to be politically dangerous and my unclear relationship with the district authorities were potential causes of trouble, and after a while he asked me to stop. Indeed, the authorities constantly monitored my presence in the field, and people whom

I met or interviewed were aware of this. As I came to recognise that such a clandestine research approach was endangering my interlocutors, I decided to go for the official, more time-consuming way of acquiring data and research permissions.

Another aspect of this situation was the secrecy surrounding the handling of most written information, including official documents, production figures, maps, statistical data, and executive decrees and directives released by central and local government bodies. Uzbekistan's agricultural sector, much like that of the European Union, is characterised by a large bureaucracy that regulates virtually all agricultural activities and thus produces many documents and much data. Information is crucial to this system and is jealously guarded. The omnipresence of the bureaucracy forced me to make continuous requests to local administrations for data and permission to access places and events. Besides putting me in a subaltern position vis-à-vis data-managing officials, these circumstances led me to interact a great deal with local bureaucrats, who first were providers of data but who also became important social actors involved in different ways in my research topic.

As a consequence, I recognised that land reform was taking place not only in the cotton fields but also in other associated places – for example, at open-air meetings or seminars (*pokaz*) called by the authorities to instruct farmers in the next steps of the agricultural cycle, in queues at government office buildings, and at the cotton ginnery at harvest time, when it was animated by hundreds of farmers delivering their cotton. All such places were stages on which the actors in agriculture played their roles. In these and similar social spaces of the agricultural bureaucracy, I could ask questions, observe interactions, catch up on rumours, and build relationships with social actors. But although it was important, bureaucracy was not the only relevant spatial dimension in my research. Returning from a pokaz I had attended with a chairman of the district branch of the Fermer and Dehqon Association, as we discussed what had been said during the meeting, he commented, 'The necessary words are said at weddings, not during work meetings'. He meant that important decisions were taken along other channels and following other criteria beyond the official façade of bureaucratic regulations and procedures.

In addressing land reform, the concept of the 'peasant world' with its social institutions, ways of life, values, duties, and strategies of defence against outsiders had some heuristic value, especially in regard to themes such as relations between peasants and administrators, whose distance or proximity often depended on (or was claimed through) kinship affiliation. Within this peasant world, weddings and anniversary celebrations to which I was invited and other, more private moments of conviviality and hospitality in which I participated were important occasions that enabled me to learn how kinship groups, families, and individuals adapted and related to the changing conditions of agriculture. In such settings, the massive consumption of alcohol surprised me; I had not come across it in urban Tashkent and Namangan, where I had been during my previous research stays in Uzbekistan. On several occasions I realised that alcohol was used to create complicity and, especially when the persons involved did not know each other well, as a way for my hosts to evade answering my uncomfortable questions.

Before I could conduct interviews on problematic aspects of resource use and the organisation of agriculture with some competence, I had to accumulate information, background data, and explanations that I lacked at the beginning of my fieldwork. Unlike other PhD students affiliated with the German-Uzbek project, I had no technical background, so gaining knowledge about the local agricultural cycle became a precondition for asking relevant questions and making informed statements to my interlocutors. Among Khorezmian farmers and officials and in the German-Uzbek project environment alike I found a marked technical approach to the perception and solving of problems in agriculture. For me these technical discourses created an accepted and uncomplicated terrain of conversation with my interviewees, most of whom were educated male heads of families, the group that almost exclusively makes up the key decision makers in agriculture. Although seldom immediately relevant for my research, such discourses were instrumental for trust-building. Land reform was perceived to be an odd topic for an anthropological investigation. Some interlocutors were puzzled about the nature of my work and thought I could be a spy.

Despite local administrators' fear, control, and restrictions on my work, I found most of them to be the human face of the authoritarian state. Far from being uncooperative, the average administrator tried to be helpful to me. At the same time, he looked for ways to safeguard his position in the administrative hierarchy, because too much initiative would expose him to attacks from his superiors. Thanks to these people's support, by the end of my research I had been able to access most of the documentation I asked for without falling into clandestine or unethical methods. Strategic considerations, an understanding of the rules and nature of the agricultural hierarchy, and 'patient deference' (Zanca 2000: 158) were the preconditions for these results. The following entry from my field protocols illustrates the inevitable ambiguity resulting from this sort of 'data-hunting' approach:

> I wanted to talk to Ismoil, but not in Komiljon's presence, since I thought the questions about land reallocation I wanted to ask Ismoil would be too delicate, thus unacceptable for Komiljon, given his high position. But as is often the case, once you are before the big

man you cannot hide anymore, and I also don't know Ismoil well enough to count on his complicity. ...

Agriculture in Yangibozor is happening all around one long street, plus the *hokimiyat* [district administration] building, as all the main office buildings in the surroundings are there: banks, *boshqarma, uyushma, prokuratura,* Raistat[26], post office, *militsiya*. All this makes it very visible. There is no equivalent to the piazza of our Mediterranean towns in Khorezmian culture, but this area is almost one. If you go to look for a person there, you cannot have a confidential reason to do so. The very reason why you are on this 'parquet' must be officially presentable. At any time a *yoshulli* [boss] could come along, which would put your fellow in an embarrassing position, because you should have asked his superior and not him.

My struggles during fieldwork mirrored the everyday difficulties faced by my Khorezmian companions. My methodological concerns and their existential concerns both resulted from attempts to cope with a state able to channel citizens' access to scarce resources but unable to offer enough resources to satisfy legitimate demands, a situation formulated in the paradox of the 'strong-weak state' (McMann 2004).

I found that behind people's apparent solidarity and amity, the social climate surrounding agriculture was marked by mistrust, enmity, and intrigue. At weddings, where people might show up by virtue of a common workplace despite open hostility with the host family, or at the district agricultural headquarters, where people stood aside to discuss matters away from their colleagues, such patterns were ubiquitous, despite people's maintenance of a façade of peaceful relations. In the everyday management of ambiguous social relations, Khorezmian rural actors were no exception to what Torsello, writing about postsocialist Slovakia, called the 'strategic use of mistrust', understood as 'doing one thing and saying something different' (Torsello 2003: 220).

My access to this kind of perspective on local relations was made possible by the confidence I gained over time on the part of the staff of a local administrative unit concerned with the farmers' movement in the district. My project colleague Nodir Djanibekov introduced me to the Yangibozor District Fermer and Dehqon Association (FDA) at the beginning of my fieldwork in 2003 – a few months after the district, ahead of others of Khorezm, had accomplished the complete turnover of collective land to newly installed private farms. Djanibekov remarked that some friendly officials in the asso-

---

[26] The *boshqarma, uyushma, prokuratura,* and Raistat were, respectively, the offices of the district agricultural department, the Fermer and Dehqon Association, the prosecutor's office, and the district statistical committee.

ciation would 'speak frankly' about the privatisation. Over time the FDA proved to be a good vantage from which to observe evolving agricultural policies. Until the end of my fieldwork I followed the vicissitudes of this association, its frequent changes of staff, and its changing relationships with farmers and government bodies. At a district level, my closeness to its personnel enabled me to access many farmers and notables of local agriculture in an uncomplicated way, when I was introduced by a staff member. In contrast, 'external' approaches and official requests for data in the domain of more powerful and less accessible governmental bodies succeeded only partially, if at all.

Given the political sensitivity of my research topic, my status as a researcher, even if one privileged by affiliation with a United Nations organisation, remained problematic until the end. Land distribution and ownership, state interference in agriculture, cropping patterns, the web of governmental and semi-governmental organisations involved in agricultural policy, perceptions of changes in agriculture, and even local issues such as labour processes in the field and the organisation and income strategies of farms proved to be surprisingly sensitive and therefore difficult to research straightforwardly. Most of my relevant insights into the agricultural system emerged unexpectedly, by my just 'being there' and following people in their everyday affairs.

Yangibozor

I chose to concentrate on Yangibozor, a district (Russ. *rayon,* Uzb. *tuman*) in Khorezm region, because decollectivisation had already been implemented throughout the district by January 2003; it acted as a trail blazer for the reforms. While decollectivisation was being applied more gradually in other districts, usually involving one or two former collective enterprises, Yangibozor was one of the four districts in Uzbekistan in which the government first introduced decollectivisation district-wide.

Yangibozor is one of the smaller and more recently created districts of Khorezm. Because of its location on the left bank of the Amudaryo, its water supply for irrigation is richer than that of districts farther downstream along the main irrigation channels. Soil salinity and water shortages are less a threat than in the 'inner' districts, which lie farther from the primary water source. With its agricultural production focused on cotton, wheat, and rice, and with almost no trade, manufacturing, or industry facilities except for the cotton ginnery in the district centre, Yangibozor at the time of my research was truly a rural district, representative of large parts of the landscape in Khorezm and in Uzbekistan.

Before the 1980s, what is now Yangibozor was divided among the neighbouring districts of Shovot, Urganch, and Gurlan, three historical centres of the former khanate of Khiva and sites of important bazaars, which the people of Yangibozor still visit to sell their agricultural surpluses and to buy commodities. The Yangibozorli saying that 'wealth finds its way to Gurlan or Urganch' (and does not remain in the district) reflects the importance of trade, in addition to agriculture, as a source of wealth in rural areas. Ironically, Yangibozor means 'new bazaar'. Unlike the villages of the area, some of which are very ancient, the district centre that gives the newly created district its name was established during the Soviet period, after a reorganisation of the territory aimed at redefining administrative boundaries in relation to the growing population. Because of its strong neighbours, Yangibozor remained small. The town's bazaar is less important than those of the neighbouring cities, and Yangibozor today is considered – if such a ranking is possible – to be among the less developed districts in Khorezm region.

In 2004 one could not say that decollectivisation had brought better conditions of life to rural communities. Living conditions in Yangibozor still reflected the decay and stagnation resulting from what Kandiyoti (2003a: 251) referred to as 'demonetization and reagrarianization'. The devaluation of wages, delayed payments of salaries in the public sector, lack of cash, pauperisation, and increased dependency on agricultural produce for securing one's livelihood were still as topical as they were in the early years of independence. Many of the symptoms of the postsocialist crisis had lost their exceptionality and had slowly become the normal state of affairs, and Yangibozor remained predominantly rural and poor.

In May 2004 a district officer estimated that among Yangibozor's approximately 65,000 inhabitants, spread out over eight village council jurisdictions and the district centre, approximately 32,000 were of working age, and 24,000 of them worked directly or indirectly on the roughly 19,500 hectares of irrigated land that were in use for agriculture in the district. Of those 24,000, 18,000 were actually 'in the fields', as this man put it, while the remaining 6,000 worked 'in trade, construction, and other services', stepping in only when their labour was most needed, 'during the cotton harvest'.[27] Factories of the Soviet period had closed in the early years of independence. With the exception of the district cotton ginnery, which seasonally employed 200 to 300 workers during the harvest, and a few small brick factories, there were no other industries in Yangibozor in 2004.

---

[27] Although this estimate seems realistic, it neglects labour migration, which seems to play an important role in the district, as I discuss later.

The early years after independence were characterised by waves of privatisation of collectively owned buildings, machines, and other assets. At the same time, in many public administrations and schools, salaries were being paid not in cash but in kind, through a system of vouchers that enabled recipients to obtain certain goods at state-owned retail shops. At the time of my research, most public employees in the district were being paid in cotton oil, sugar, and wheat flour. Part of their salaries were withheld in payment for electricity. Public employees entered business and agriculture driven by the need replace their vanishing income. Unemployment was widespread, and most people made their living through activities they did not consider to be 'real work', meaning that salaries had ceased to be the most important source of income. In almost every family in Yangibozor, one or two men were abroad in Russia or Kazakhstan for seasonal work, mostly in agriculture and construction. As in other parts of Uzbekistan, arrangements regarding waged labour, domestic work, and family provisioning established during the Soviet period could not be maintained, and people had to adapt by 'retreating' into petty trade and individually manageable social arrangements (Yurkova 2004: 175). In rural districts such as Yangibozor, when the collective farms were no longer paying salaries in the 1990s, sharecropping emerged as the substitute form of remuneration.

In Yangibozor today, cotton and wheat are the most important crops. Much importance is also given to rice, which farmers and peasants in Khorezm prefer because of its profitability. Irrigation water is received at a negligible cost, and rice can be sold in the bazaar for good money, so it is considered the main cash crop for peasants and farmers, just as cotton is for the government. For these reasons, entitlements to land and cropping rights, as well as privatisation and the implementation of reforms in general, are perceived to be sensitive political issues in Yangibozor. For people in agriculture, entitlements are the desirable 'cake' over which to compete.

A statistical document from the district land-measurement department for 2006, the most complete recent document I could access on agricultural production in the district, summarised quotas for the allocation of land to different crops. It showed that 19,293 hectares fell under the category of main agricultural land (*ekin yerlari*). Within that category, land was planted as follows (for percentages, see chart 1): 12,000 hectares in cotton, 3,716 hectares in wheat, 1,643 hectares in forage crops, 1,280 hectares in rice, 283.5 hectares in vegetables, 140 hectares in melons and gourds, 65 hectares in potatoes, and 48.6 hectares in barley. According to this source, 1,634 hectares were allocated to households as subsidiary smallplots (*qo'shimcha tomorqa*).[28]

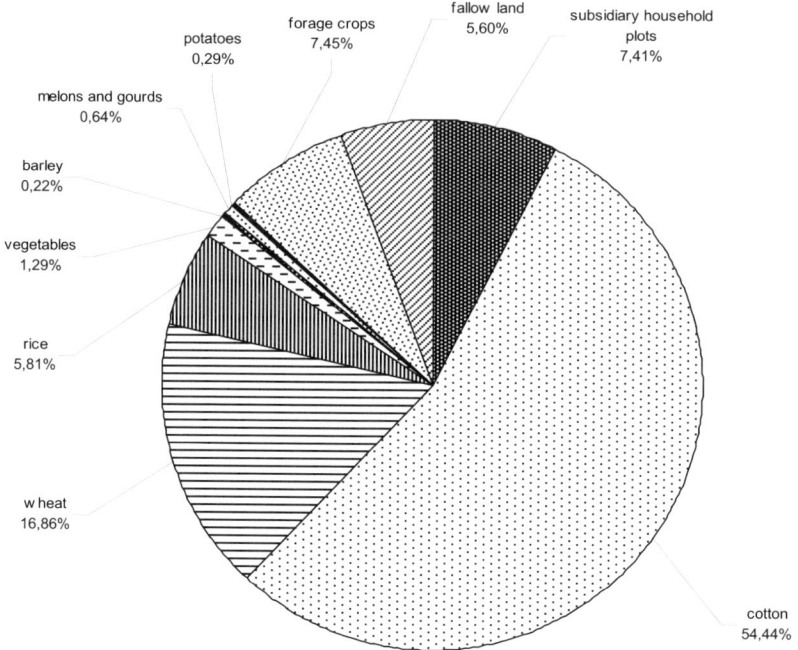

Chart 1. Land allocation in Yangibozor district, 2006. Source: author's fieldwork data, 2006, based on district land measurement office data.

Fieldwork

The research methods I adopted during fieldwork consisted mainly of participant observation as understood by Bernard (1995: 136ff.), of semi-structured and unstructured interviews and informal conversation (Bernard 1995: 209ff.), and of other time-proven methods in the repertoire of socio-anthropological craftsmanship (Gluckman 1967).[29]

---

[28] Imprecision of data can be attributed partly to the accounting system left by the planned economy. Still in use, it distinguishes between a planned (at the beginning of the year), an intermediate (in June, declared on form SX-9), and a final (effective) figure for every item of economic data in agriculture. The imprecision also reflects local fluctuations in the planning of agricultural activities, although a roughly steady cropping pattern was maintained every year.

[29] During fieldwork, I opted for a selective application of Elwert's (2003: 29ff.) 'bundle of methods'. Note-taking was the principal form of data collection. I put notes down in protocols subdivided into paragraphs with multiple keywords, finally organised in keyword-sensitive databases. Besides this, I obtained or produced statistical data, documents on laws and

I carried out my research during three periods. In a preliminary research trip from the end of March to the beginning of June 2003, I aimed at positioning my research within the framework of the ZEF/UNESCO project and Khorezm. My second stay consisted of nine months in the Khorezm region, from January to October 2004, at which time my most important activities were field and village visits, interaction with local officials, requests for statistical data, work on archival data sources, and preparation and implementation of a household survey. In autumn 2006 I returned to Khorezm for another month. This third visit proved useful in seeing how the situation had evolved during my absence.

My plan for carrying out a semi-structured household survey, solicited by the project during my longer stay, was designed to introduce myself in the research area, acquaint myself with villages throughout the region in order to select village case studies later on, produce a statistically representative snapshot of social stratification, address perceptions of agricultural reform among different segments of the population, and assess the importance of family networks in agricultural production. I planned to survey a randomly selected sample of up to 100 households. I decided to interrupt the survey, however, after just 16 households, because I recognised that the answers I sought depended on informants' cooperation and trust and could not be adequately supplied by a household survey. Although apparently a failure, the survey ended up being a useful experiment that indirectly provided valuable information on the reality of the rural villages. While I refined my questionnaire and interacted with members of communities and authorities, my understanding of the rural context improved. Deepening relationships with my gradually growing network of interlocutors proved in the end to be a more efficient means of achieving insights.[30]

Various local, district, and regional organisations in Khorezm are involved in producing, storing, and monitoring data relevant for a study of reforms of the agricultural system.[31] Consequently, meticulous data production exists for virtually all aspects of agricultural production, management,

---

regulations, maps, organisation plans, photographs, and tape-recorded interviews with key informants. In this book I use pseudonyms for names and places in cases for which it seemed important to protect respondents' anonymity.

[30] I find many parallels between my methodological challenges (and ways of circumventing them) and those described by Allina-Pisano for her research in the Russian and Ukrainian Black Earth region, including an 'ill-fated survey' (Allina-Pisano 2008: xxii).

[31] A few of the more important organisations were the statistical committees (Raistat, Oblstat), the neighbourhood committees (*mahallakom, elatkom*), the district agricultural department (*boshqarma*, also called Agroprom), the Fermer and Dehqon Association (*uyushma*), the prosecutor's office (*prokuratura*), banks (*paxtabank, g'allabank*, etc.), and the cotton ginnery (*paxtazavod*).

and planning. As important as these data are for a study of land reform, they are complicated to access and problematic to interpret. At the beginning of my work I looked for statistical data in order to orient my selection of research sites. But some of the data I obtained, with great difficulty, from the statistics departments were approximate, inaccurate, or aggregated in ways that made them useless for analytical purposes, so I decided not to rely on them for further planning of fieldwork. On several occasions, people raised doubts about the reliability of official statistics.[32]

There were other data sets, however, whose reliability I had no reason to doubt.[33] I made a distinction between data intended for external use (that is, requested from outside the command administrative system), which were usually of poor quality, and data for internal use, which were accurate but kept secret. The latter were the instruments for steering, controlling, and monitoring the economic activities of the districts. At the local level, they consisted mostly of long lists of single cases that required a great degree of contextual knowledge to be interpreted correctly. In Khorezm, at least since the cotton scandal of the 1980s (a result of data manipulation at various levels), data production and data management are sensitive issues. I therefore dedicated much attention to questions about what kinds of data were used, created, and managed at the district level and below, how they were used, and why.[34]

As my understanding of the local social context and my confidence in speaking the Khorezmian dialect grew, I focused more on field experiments initiated jointly with my project colleagues. My other prevailing activities

---

[32] During a party (*ziyofat*) with some friends in Yangibozor, a *revizor* (a chartered accountant for public and private enterprises) said, 'In our statistics they say that there are 25 million Uzbeks. In reality there are many more, they are 30 million or so, but they decrease the number in order not to appear as too underdeveloped a country'. During an international conference in London, a participant who happened to be a former governor (*hokim*) of Bog'ot district in Khorezm told me that I should not trust official statistics, because the figures were often embellished according to top-down, mandated prescriptions.

[33] Despite skepticism towards figures that were often said to be doctored in order to hide water waste and unequal access, I find confirmation for my statement in Veldwisch (2008: 52), who, working on water allocation and relying on similar local-level data, cross-checked them and confirmed their reliability. In my case cross-checking was, for obvious reasons, more difficult.

[34] A general premise is that aggregated data produced by the statistical committees are less reliable and more exposed to manipulation, in that they are made to fit given target figures. Unaggregated data can be more interesting, although they are often secreted and thus difficult to access. Legislation on the use of statistical data prohibits the dissemination of unaggregated data (O'zbekiston Respublikasining Qonuni, Davlat Statistikasi to'grisida, 12 December 2002). Working data produced by the agricultural organisations themselves are less exposed to this risk but are not entirely secure from manipulation. In analysing data, I tried to critically evaluate the collected material in every case, as I discuss further in chapter 3.

included participant observation and interaction in the everyday activities of farmers and local-level officials (the 'follow-your-key-informants' method) and the mapping of land redistribution in selected former kolkhozes. The latter was made possible by cooperation with a district-level land-measurement office. In exchange for digitalising its cadastre maps, which I could offer thanks to the project's equipment in Urganch, I was able to interact with land measurers in the villages of Yangibozor, who helped me compile maps of land redistribution.

Plate 2. The land measurers of Yangibozor district compiling maps of district land-use patterns at the request of the regional administration.

During most of the time I spent in Khorezm I lived in the project dormitory, a large, two-storey family house hosting up to a dozen local and international students affiliated with the project. Situated in a neighbourhood on the outskirts of Urganch, it was rented by the project from a wealthy local businessman. Neighbours referred to it as the 'hotel' (*gostinica*), emphasising our separateness from the surrounding social relations. This in some sense contradicted the tacit imperative of the classic anthropological village perspective, although it fitted the expectations of project colleagues and coordinators, local authorities, and informants and friends from nearby villages of what my proper place in Khorezm would be. Because Yangibozor was only 20 kilometres away and I had a small car and a driver at my disposal, it proved more convenient to schedule daily trips there, sometimes with stays overnight or for a few days at friends' places, than to locate myself in a village in Yangibozor.

Clearly, the low-profile, informal ways of accessing 'the field' that are usually associated with anthropology clashed with the reality of the well-equipped, highly organised project with which I was associated. Farmers, colleagues, and friends from the district who sometimes visited me at the project building in Urganch were amazed by its neo-Uzbek architecture and the equipment imported from Germany. Once acknowledged and digested, the divide between the project and 'the rest' was not as insurmountable as the anthropologist's sometimes naïve longing for an intimate insider role might posit. The good reputation the project enjoyed among many local agricultural actors helped give my research visibility and acceptance in the eyes of those with whom I interacted, and it conferred credibility on my effort to 'study up'. With its staff, students, and affiliated university teachers, the project was itself a 'Khorezmian family' with a wide-ranging network that in various ways helped the minority of foreigners among us to familiarise ourselves with topics and issues. In short, the project significantly influenced my work. Although not unproblematic, it became part of the 'workable normality' of the fieldwork interactions I managed to create over time.

During the course of my stay, a spatial focus evolved in my fieldwork. Whereas at the beginning I conducted interviews in different districts, in the end I focused only on Yangibozor, which I used as my case study. In 2003 and during the survey in the first months of 2004, I carried out all activities together with a research assistant (de facto an interpreter who helped me understand the Khorezmian dialect), but over time I managed to conduct interviews alone. The latter fact augmented my autonomy and flexibility in the research.

Chapter 2
Khorezm Region

Situated along the lower course of the Amudaryo, Khorezm is a large, irrigated oasis surrounded by the deserts Qoraqum in the south and Qizilqum in the north. It was also known as the khanate of Khiva until the Bolsheviks put an end to the tsarist protectorate with the proclamation of the Khorezm People's Soviet Republic in 1920. Soviet national delimitation of Central Asia (1924–1936) split the former khanate into what are now the northern Turkmenistan province of Dashog'uz, the Autonomous Republic of Karakalpakistan, and the western Uzbek region (*viloyat*) of Khorezm, where the former capital of the khanate, Khiva, is located.[35] Established in 1938 as an oblast of the Uzbek SSR, Khorezm today is the smallest of the 12 viloyats of the Republic of Uzbekistan, extending over 6,100 square kilometres. It is a predominantly rural region with a fast-growing population, increasing from 688,500 inhabitants in 1978 to nearly 1.4 million in 2004, of whom only 321,200 were urban.[36] Administratively, Khorezm is subdivided into 12 districts, called *rayon*s in Russian and *tuman*s in Uzbek, as shown on map 1.

---

[35] For a concise description of the USSR's policies of national delimitation, see Fierman (1991: 16–18). Concerning the delimiting of the Soviet Khorezm region, Sabol (1995: 235–236) argued that Khorezm Uzbeks at first 'had been only mildly interested in the project' of their own nation but then 'suddenly proposed a larger, federated Khorezm'. However, the 'Bolsheviks were able to suppress these demands and proceed with their plan'. Although the Soviet national delimitation of Central Asia is often presented as a *divide et impera* policy imposed from above, historical research (Eisner 1994: 110) has highlighted the role played by local bureaucrats in conflicts over territories (see also Fedtke 2007 for the case of Bukhara).

[36] Data for 1978 are from *O'zbek Sovyet Entsiklopediyasi* (1979), and those for 2004, from the Khorezm Oblstat, or statistical committee.

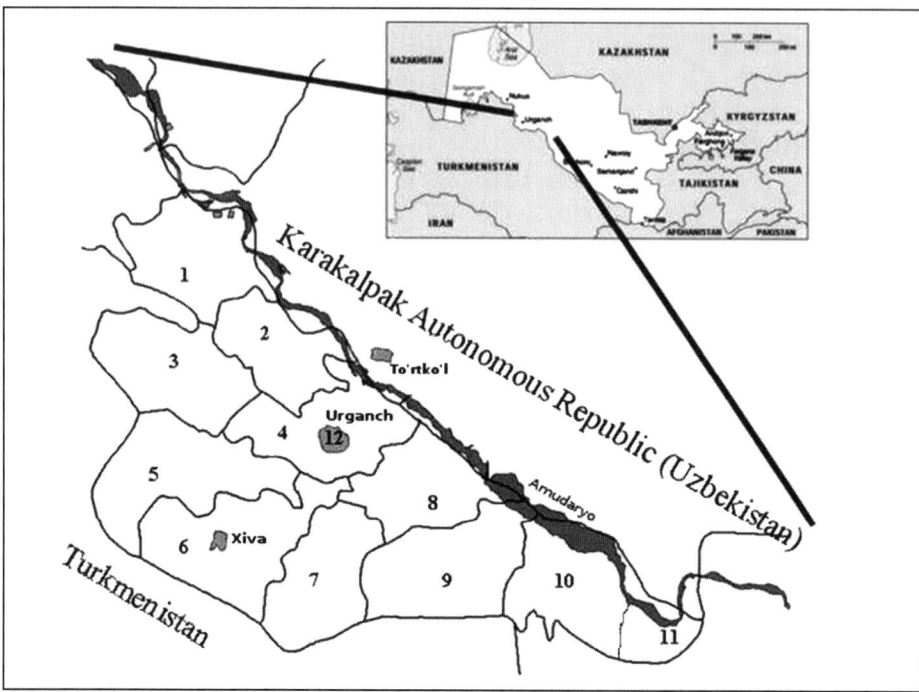

Map 1. The region of Khorezm in Uzbekistan. Districts are numbered as follows: 1, Gurlan; 2, Yangibozor; 3, Shovot; 4, Urganch; 5, Qo'shko'pir; 6, Khiva (Xiva); 7, Yangiariq; 8, Xonqa; 9, Bog'ot; 10, Xazorasp; 11, Pitnak; 12, Urganch city (Urganch *shahri*). Adapted from ZEF/UNESCO Khorezm project.

Although agriculture's share of regional GDP is declining, it still plays a predominant role in Khorezm's overall economy.[37] Because the state maintains a mandatory crop policy, cotton and winter wheat are the main crops produced in Khorezm. According to Kienzler (2010: 9), 7 to 8 per cent of cotton and 4 to 5 per cent of winter wheat produced in Uzbekistan in 2004 originated in Khorezm, covering, respectively, 45 per cent and 21 per cent of the 275,000 hectares of irrigated agricultural land in the viloyat. Rice is the third most important crop, and other important produce is fruits such as apples, apricots, melons, and grapes and vegetables such as onions and carrots. Unlike cotton and wheat, these minor crops in Khorezm are grown mainly on small household plots. Among the other important agricultural activities, cattle breeding plays a special role. In the past the collectives had large herds of cattle, but during my stay in Khorezm breeding was done almost exclusively by households.

---

[37] According to Djanibekov (2008a), agriculture accounted for two-thirds of regional GDP in 2003 but only 45 per cent in 2005.

Other sectors also contribute significantly to the region's economy, although their contributions are difficult to estimate. Among them, remittances from labour migrants in Tashkent and, especially, abroad, together with trade (and smuggling) across the Turkmenistan border, have grown in importance.[38] Since independence, the tourist sites of Khiva, which UNESCO named a World Heritage site, have attracted a growing number of international visitors and promoted the region's image in the world. Meanwhile, the government is pushing the expansion of local industry, which had collapsed and virtually disappeared in the aftermath of the disintegration of the Soviet Union. During the time of my fieldwork in Khorezm, the government's plans were still in their infancy, and the only relatively large factory that was still running in Urganch was the distillery (*vinzavod*). Among the smaller enterprises, a number of cotton processing plants, a large mill in Xonqa, and minor food-processing factories were mostly leftovers from Soviet industrialisation. A carpet textile factory, a joint venture involving foreign capital, had recently been opened in Khiva *rayon*.

As one of the most densely populated and intensively irrigated cotton-growing areas of Uzbekistan, Khorezm had population growth (3.9 per cent annually) and population density (190.5 inhabitants per square kilometre) well above the national average.[39] Of the more than 1.3 million inhabitants of Khorezm, 'about 70% are rural, and about 27.5% live below the poverty line of 1US$ per day' (Wehrheim and Martius 2008: 6). Although these figures make Khorezm look similar to less-developed, third world countries, the comparison is misleading. Under the Soviet Union, Uzbekistan's rural population became accustomed to the high living standards that came with a centrally commanded, modern agro-industrial sector specialising in cotton (Khan and Ghai 1979: 40–63; Patnaik 1996; Tursunov 1996).[40] With education, health, and living standards rapidly declining during the 1990s, 'villagers feel they are in the grip of de-modernization' (Humphrey 1998: xi), as is the case in many other rural parts of the former Soviet Union.

Uzbeks make up the overwhelming majority of Khorezm's population. Rural areas especially are characterised by ethnic homogeneity and are

---

[38] During my fieldwork in Urganch, I gained the impression that most drivers bought their petrol on a street that had developed into an unofficial marketplace for petrol (*neft bozori*). Smuggled in from Turkmenistan and displayed in plastic bottles at the edge of the street, this petrol was much cheaper than that sold in the official petrol stations. The diesel necessary for tractors and in agriculture was unavailable in such bazaars. Other goods brought in from Turkmenistan were electronic equipment, stereos, and televisions, which apparently were cheaper there than in Uzbekistan.

[39] Data are for 1999 and are taken from Defrade (2000: 167).

[40] For a discussion of post-war agricultural and social development in Khorezm, see Matniyozov (1997: 166ff.).

overwhelmingly populated by Uzbeks. Khorezm's countryside maintained its ethnic homogeneity throughout the Soviet period, except in a few former sovkhozes and kolkhozes where non-Uzbek minorities lived. Some Koreans, for example, were sent to help develop the rice-growing sovkhozes near the Amudaryo in the 1970s, and sometimes women from other republics of the USSR moved, after marriage, to the villages of their husbands, whom they had met during the men's military service with the Red Army. According to the *O'zbek Sovyet Entsiklopediyasi* (1979: 366), in 1970, 92.1 per cent of the population of Khorezm consisted of Uzbeks. Karakalpaks, Turkmens, and Russians made up 2.4 per cent of the population; Kazakhs, 1.6 per cent; Tatars 1.3 per cent; and Koreans, 1.2 per cent. According to a Korean acquaintance of mine from Urganch, the mid- to late 1990s saw the non-Uzbek, 'European' population, especially Russians, leave the region en masse. Non-Uzbeks, especially those from rural areas, who had a chance to leave did so. The ethnic composition of Khorezm, like that of other regions of Uzbekistan, follows the pattern of a predominantly, if not almost entirely, Uzbek countryside and more mixed populations in urban areas, where Russian speakers and Uzbek speakers live in separate neighbourhoods.[41]

Khorezmian Uzbeks, however, retain a sense of distinctiveness that makes them different from other Uzbeks, both by self-perception and by the ascription of others. Although Khorezmians are not recognized as a separate ethnic group, they maintain a distinct sense of belonging to their region, in opposition, for example, to Uzbeks of the Tashkent, Bukhara, and Samarkand regions or from the Ferghana Valley. In Uzbekistan, Khorezmian people are considered to be Uzbeks with strong peculiarities of language and culture that are rooted in their belonging to a political entity that existed continuously for centuries, until 1924 (Baldauf 1995: 40–41).

Khorezmian Uzbeks

According to Sazonova (1952) and Tolstov (2005 [1948]), the people living in the southern part of the Khorezm oasis, including the population of today's Khorezm region, along with the people of the area around Dashog'uz in Turkmenistan and, across the Amudaryo, in the neighbouring tuman of To'rtko'l in Karakalpakistan, are the most direct descendants of the ancient

---

[41] Rural ethnic homogeneity can be seen in Yangibozor in, for instance, the village of Bog'olon, where 9,856 Uzbeks, 18 Tatars, 14 Kazakhs, 8 Bashkirians, 5 Russians, 3 Turkmens, and 1 Moldavian were registered in 2004, according to the Bog'olon village booklet (*Xorazm viloyat Yangibozor tumani Bog'olon qishlog'i pasporti*) published that year. The village of Qalandardo'rman had 6,053 Uzbeks, 79 Kazakhs, 25 Karakalpaks, 11 Tatars, 5 Bashkirians, 2 Russians, and 1 Tajik registered in 2004 (*Xorazm viloyat Yangibozor tumani Qalandardo'rman qishlog'i pasporti,* 2004).

Khorezmians. During the late Middle Ages this sedentary, essentially still Iranic-speaking population of agriculturalists and craftsmen mixed with the Turkic-speaking Uzbeks of the Golden Horde, who in turn were closely linked to the peoples of the eastern and southern Priaral region (the semi-nomadic Karakalpaks and the nomadic Kazaks).

The original inhabitants of the oasis exerted a strong influence over the economy and culture of the Uzbeks living among them, who gave up their nomadic lifestyle and tribal affiliation, became sedentary, and took up irrigated agriculture. At the same time, the autochthonous inhabitants gave up their original denomination, 'Sart', and merged with the Uzbeks (Sazonova 1952: 248).[42] Although the two terms were still commonly used at the beginning of the tsarist period in Central Asia, the term *Sart* had started to become obsolete even before the Bolshevik revolution. The mixing of these two groups, especially in today's Khorezm region, was favoured by a long period of coexistence within the boundaries of the unified khanate of Khiva, as well as through intensive cultural and economic links.

Nevertheless, some of these linguistic differences live on today. In Yangibozor, an immaterial line divides the district into halves, and the language spoken in the southern half is related to the dialect (*sheva*) of the former 'Khiva Sarts' (Baskakov 1960). In the north, towards Gurlan, some people identified themselves as 'Qipchaq' and spoke a peculiar dialect. As Kehl-Bodrogi (2008: 49) experienced, dialects in Khorezm can vary considerably, especially between the southern and northern districts. Occasionally these language differences can translate into even more meaningful differences, as when long-dormant prejudices re-surface and stereotypes are revived in one district's dislike for another. One example is the diffidence of the people of Yangibozor towards the 'greedy and treacherous' (that is, 'Sart-like') people of Xonqa.

After the national delimitation, in the Uzbek SSR the Khorezmians, along with Turkic populations in the other oases of former Turkestan and the emirate of Bukhara, were categorised as belonging to the Uzbek ethnic group, although this name was in fact applied to a variegated population

---

[42] Sazonova's description oversimplifies a complex historiography. The name Sart was, strictly speaking, neither an ethnic term, since it could refer to both Uzbeks and Tajiks all over the territory of former Turkestan, nor a reference to a particular language (the sedentary city dwellers spoke Uzbek, Farsi, or both). For this reason and because of its derogatory connotation, it quickly became obsolete after the Soviets, in their nationalities policy, developed ethno-linguistic definitions of Central Asia's populations. By the time of the 1926 census, Sarts were no longer among the listed nationalities. See Subtelny (1994: 49–52). For Abashin (2007), on the contrary, the term Sart was intimately linked to a mode of ethnographic classification resting on a 'Tsarist imperial vision', which the Bolshevik takeover promptly rendered obsolete.

encompassing different ethno-historical and language backgrounds.[43] The Khorezmian dialect, the language spoken by Uzbeks of the oasis today, originated in the Oghuz dialect, a branch of the Turkic languages different from the Ferghana dialect, which in the 1930s was chosen to become the standard, literary Uzbek language (see Allworth 1990). As described by Baldauf (1995), in Central Asia's traditionally multilingual and multiethnic context, forms of social identification rooted in common territorial origins always overshadowed the importance of other identity markers, such as affiliation based on ethnic descent or language. This is especially true for Khorezm, and so Khorezmian 'particularism', expressed in language, sense of belonging, and cultural practices, lives on vigorously today (Kehl-Bodrogi 2008; Turaeva 2008).

A short verse from a popular poem, which I translate freely, expresses Khorezmians' pride in their country and language: 'I have a people that is looked after by God / My language is a delicious dialect / In my blood there is knowledge and craftsmanship / In my fatherland, Khorezm, / we have a *to'y*, a celebration, every day'.[44]

I heard this verse recited during a *to'y* I attended for the sixtieth birthday of a former kolkhoz chairman (*rais*). Besides national pride, it expresses Khorezmians' love for such celebrations and for *tomosha*, the enjoyment of the spectacle, which was more pronounced in Khorezm, where *to'y*s assumed extravagant forms, than in the *to'y*s I observed in the Ferghana Valley. National holidays such as Independence Day, or Navro'z, are celebrated in a 'Khorezm-like' (*xorazmcha*) way. During Navro'z 2004, presided over by the district governor and other city dignitaries, the stadium in Urganch hosted the traditional fights held to salute the beginning of spring – first fights between rams, then between wrestlers (*polvon*), and finally between dogs, all held as entertainment. The city mayor handed out prizes to the winners, to the applause of a large crowd. Occasions to express pride in the qualities and distinctiveness of the Khorezmian dialect, music (especially the Khorezmian *lazgi*), dances, and dishes such as Khorezmian *palov* and *yumurtabäräk* were seldom missed. Many Khorezmian Uzbeks I met during my fieldwork emphasised their attachment to their native Khorezm rather than to their Uzbek identity, professing, 'I am Khorezmian [Xorazmlikman]'.

---

[43] For an extensive problematisation of the term *Uzbek* as an ethnic or identity category, see Schoeberlein-Engel (1994: 46–72) and Finke (2006).

[44] In Uzbek: *Xudo qaragan elim bor / Mazali sheva tilim bor / Qonimda hunar-ilim bor / Ota vatanim Xorazm / Har kun to'y, har kun bazm.*

Plate 3. A wrestling match held during the Navro'z celebration, Urganch, March 2004. Photograph courtesy of Kirsten Kienzler.

Kehl-Bodrogi (2008: 47) noticed the way many Khorezmians take a reluctant stance towards official historical narratives and especially towards the official 'cult of Timur'. Statues of Timur (Tamerlane) can normally be found in all major Uzbek cities, given his symbolic function in Uzbekistan's national ideology (Ilkhamov 2007). Having invaded and ravaged Khorezm several times, Timur is detested by Khorezmians, and Urganch is the only regional capital in Uzbekistan where the statue of a Khorezmian statesman, Jaloliddin Manguberdi, takes the symbolic place usually occupied by Timur.[45] Once, during a banquet at a farmer's house in Yangibozor, an inebriated attendee said he was 'against Timur, who had destroyed Khorezm many times, no matter what the people say in Tashkent'. Given the presence of a foreigner and of a guest from Tashkent, the host immediately stopped the man and asked him not to talk about politics. Fear and caution characterise any talk about politics in the presence of outsiders, but all the more so when it takes up the banner of regionalism. I agree with Kehl-Bodrogi that

---

[45] Following the Mongol invasion of 1220, Jaloliddin Manguberdi became the last ruler of the Khwarezmid empire, after his father, the last Khorezmshah, was killed. He himself was assassinated in 1231. The people of Khorezm still consider him a 'national' hero.

political antagonism against Tashkent is held in check in Khorezm and that neither a politicised Khorezmian nationalism nor a desire for secession from the rest of the country has developed. Relations with the political centre, however, are a sensitive issue, and not only for reasons rooted in differences of identity.

Rather, asymmetries of power between a backward, peripheral province and the political centre have sealed the region's dependency and subalternity vis-à-vis Tashkent. As I discuss later, cotton plays an important role in defining these relations. An acquaintance who had worked for the district administration in Yangibozor once told me that to him, Tashkent 'materialised' in Khorezm's world of agriculture in the form of either delegates (*vakil*) from the ministry who periodically showed up to monitor cotton production or scientific delegations from research institutes in the capital who came to introduce novelties in the way agriculture was done locally. In his words, both groups would routinely 'go mad' (*jilli bo'p ketadi*) when confronted with the inefficiencies and obstacles that local conditions posed to the implementation of centrally commanded directives. At times, backwardness and underdevelopment could become shields that screened external intrusion.

Regarding the past, Zadykhina (1952: 281) wrote that in 'the pre-revolutionary period the Khorezmian Oblast was the most backward, neglected region of colonial Turkestan. The isolation of Khorezm, the absence of convenient means of communication within it, the despotic form of government, the oppressive rule of the beys and the priesthood, cruelly exploiting the workers, and the colonizing policy of the Tsarist government contributed to the preservation ... of the backward feudal forms of society right up to the Great October Socialist Revolution' (quoted in Battersby 1969: 17).[46] However ideologically biased, these words highlight the region's geographical isolation at the time. According to Baldauf (1995: 41), Khorezm traditionally was 'self-oriented', and in its contacts with the outside world it privileged the north-west (Astrakhan) over the south-east (Bukhara and Tashkent). Today's boundaries and stricter visa regime with neighbouring Turkmenistan further deepen the locally perceived sense of isolation.[47]

The divide between Khorezm and more developed regions of Uzbekistan evoked by Zadykhina lives on today, although in less crude forms.

---

[46] Zadykhina (1952: 281) noted that 'in comparison with other oblasts there is a large number of camels here, mostly for pack transport of goods'.

[47] At the time of my fieldwork there were three daily flight connections to Tashkent and a weekly flight to Moscow from Urganch airport. The railway and motorway that link Khorezm to the Bukhara region crosses several hundred kilometres of desert.

Khorezm has a rich cultural tradition and a highly productive agricultural sector, yet the region's image in the country is still overshadowed by the cliché of its being peripheral, backward, and somewhat different from the other, more populous and economically capable viloyats in the east and centre of the country. Before I began my fieldwork, a friend from Tashkent warned me that it was 'difficult to gain Khorezmian people's confidence', although he added that 'if you have gained their confidence, you are their friend for life'. I heard other such anecdotes but came across no written literature in which anyone attempted to reflect more thoroughly on the uniqueness of present-day Khorezm.[48]

Any attempt to find a 'fixed' regional character and distinctiveness risks reducing the complexity of social relations to reified stereotypes, especially in regard to domains as fluid as culture and politics. Yet comparing my experiences of peasants, farmers, and local officials in Yangibozor's agricultural world with my previous knowledge, gathered in the Ferghana Valley and in Tashkent, induces me to think that Khorezm does have its peculiarities of cultural and political *habitus*. Although difficult to capture and transmit, such regional peculiarities seem to be confirmed in the colourful words of the ethnomusicologist Theodore Levin (1996: 166–167):

> Directness, plainness, intrigue with power – these are qualities that seem to run not only through Khorezm's music but also through much of its cultural life. They are qualities evident in the design of houses – huge graceless rectangles built of thick, unwhitewashed clay – and in the entertainment that seems most to captivate Khorezmian males: animal fights. One afternoon, OM and I sat for hours in a stadium with five hundred or so men and boys who watched with apparent fascination as pair after pair of sheep were sent rushing on a collision course to butt their horned skulls together with a resounding thud. OM attributed the extroverted and direct Khorezmian temperament, the worship of physical strength and power, in art as well as in life, to the effects of Khorezm's extreme continental climate: frigid winters and scorching hot summers.[49]

---

[48] Another such anecdote I heard concerned Khorezmians' apparently superior astuteness, whereby they are identified as 'foxes' in contrast to the 'dumb oxen' from other parts of the country.

[49] Levin himself saw this last statement with a more weighted sense of proportion, continuing in a footnote: 'The molding of the Khorezmian character must have been influenced by more than the effects of climate, for other, more stereotypically introverted Central Asian peoples live in similar climatic conditions' (1996: 302, note 2).

Khorezmian Communities

During my first visit to Khorezm, I was at first disoriented by the fact that many rules and institutions I knew to be essential in the lives of Uzbek communities farther east did not match the situation in Khorezm. When, at the beginning of my fieldwork, I looked for 'elements of traditional solidarity' (Elwert 1983: 203ff.) that worked as the 'glue' of social relations, I searched for the neighbourhood (*mahalla*) and the mosque, expecting them to be the two wardens of traditional morality in Khorezmian communities. Elsewhere in Uzbekistan, these two institutions stood in opposition to the new moralities and lifestyles imported during the Soviet atheistic struggle for modernisation and embodied in organisations such as the Komsomol (the union of Communist youth) and, in rural areas, the kolkhoz and the village administration (*sel'soviet*). Instead, I found mahallas and mosques largely absent in Khorezmian communities.

The mahalla, a characteristic of traditional Central Asian urban settlements, is a neighbourhood community that should be conceived of as a 'unit of social organization intermediate between the household and the village' (Rasanayagam 2002b: 84). It embodies an ideal of mutual care and assistance between neighbours and is a carrier of implicitly codified social obligations and rules of life in community. For instance, the mahalla is involved in the performance of Uzbek life-cycle rituals, so it is central in maintaining and reproducing Uzbek community life (Koroteyeva and Makarova 1998b).

What is known as a mahalla in the rest of the country is called an *elat* in Khorezm. Besides terminology, there are more substantial differences. In the mahallas, elderly people are the reference group of authority, exerting influence over the behaviour of the community. They usually meet for common prayers in the mosque or the teahouse (*choyxona*). In the council of the elders (*oqsoqollar kengashi*) they discuss and decide on community matters. In the Khorezmian elats I knew from my fieldwork, old men did not meet regularly as they did in mahallas elsewhere in Uzbekistan, either to pray together or to take joint decisions as a committee. Their influence was more circumscribed, limited to the family sphere.

In Khorezm, an elat originated as a community of related neighbours, descended from a common ancestor, although in recent years elats have been mixed up and have lost their homogeneity. Links between neighbours are looser in the elat than in the traditional mahalla. In the elat, social relations can be more anonymous, and social control weaker, than in the mahalla, where the neighbourhood committee may exert strong control over the organisation of *to'y*s such as weddings and circumcision feasts, deciding on their sizes, durations, and even menus. In Khorezm, neighbourhood committees, known as *elatkom,* interfere less in the organisation of *to'y*s, and members of kinship networks are more free to make arrangements for themselves.

Under the Soviet Union and even more strongly since independence, the government has sought to increase its proximity to the institution of the mahalla, with the vested interest of increasing its legitimacy among the population. As Koroteyeva and Makarova (1998b: 139) put it, 'the mahalla has always been a cell of society; the Uzbek state is presently attempting to make it a cell of the state as well'. By formalising the rules and roles of the mahalla, the state has attempted to manipulate its symbolic importance as a traditional, locally accepted institution for its own political goals (Massicard and Trevisani 2003). In Khorezm, because of the elat's difference from the mahalla, this policy has been less successful. Whereas in the east of the country the mahalla chief appointed by the government is called by the Uzbek term *rais,* or *oqsoqol* (meaning 'white beard', a respectful way to refer to the aged dignitaries of a community), in Khorezm the word used for the head of an elat committee (an *elatkom*) is *sho'ro*, the Uzbek word for 'Soviet', which during the Soviet period was used to denote the chief of the sel'soviet. In the dialect of Yangibozor, people refer to him as *sho'ri* or *sho'ra*, or sometimes just as the rais of the elatkom.

As elsewhere in Uzbekistan, in Khorezm mosques were closed and religious life repressed during the Soviet years. But in this respect the situation in Khorezm again differs from that in other regions of Uzbekistan. As Kehl-Bodrogi put it in the title of her 2008 book, 'religion is not so strong here'. Unlike in the more conservative areas in the centre and the east of Uzbekistan, few inhabitants of Khorezm follow the precepts of Islam strictly. Knowledge of Islamic precepts is less widespread among Khorezmians, Islamic practice is less rigid, and mores are less conservative.

Kehl-Bodrogi (2006) observed that sheikhs – guardians of the many holy shrines, tombs of saints, and cemeteries – were widely respected in Khorezm but not recognised as dominant religious and moral authorities. According to her, this meant not that the Soviet anti-religious campaigns had been successful but that the weakness of sharia-minded religiosity and Islamic orthodoxy in Khorezm made room for a local religiosity based on the veneration of shrines, the performance of life-cycle rituals, and conformity to customary practices perceived to embody tradition. Levin (1996), too, emphasised the strength of local traditions in Khorezm. Quoting a fellow musician, he wrote:

> There's still a strong predisposition toward local tradition in Khorezm that probably has to do with its location on the periphery, relatively speaking. It was on the periphery of the Islamization of Transoxania, just as, centuries later, it was on the periphery of Russi-

fication. Khorezm wasn't affected as much as Tashkent, Samarkand, or Bukhara by the Soviet "struggle against the old". If it had been the center of attention, they would have tried harder to modernize it'. But Khorezm's embrace of tradition also has a contradictory side, for there, more than anywhere else in Uzbekistan, communism found a receptive audience. As one musician put it, 'We believed. We were close to communism. We believed that you had to change people's consciousness. Moreover, religion wasn't strong here the way it was in Tashkent, Ferghana, and Bukhara. We didn't go to the mosque on Fridays (Levin 1996: 163).

During the 'struggle against the old', local traditions were strongly altered, and religious rituals that played a role in the agricultural cycle fell into disuse. In a discussion with a university teacher, a pious Muslim from Urganch, I was told that before Soviet times, in Khorezm as in the rest of Uzbekistan, it was customary to pray and to 'say Amin' before the sowing of the seeds. Some older peasants still did this, I was told, but not many. 'During the Soviet period one would get punished for such things; nobody did it if not in secret. Therefore it fell in disuse'.

In the late 1980s and early 1990s, the years of the 'revival of Islam' in Uzbekistan (Hilgers 2009: 43ff.; Khalid 2007: 116ff.), many new mosques were built and old, closed ones reopened all over the country, in what could be seen as a grassroots movement. In the following years this movement was contained by a state-led, top-down approach, with the state monopolising the construction of new mosques and banning community activism (Rasanayagam 2010). Khorezmian communities, especially rural ones, were less affected by this phenomenon than communities elsewhere in Uzbekistan. Unlike in other parts of the country, mosques did not mushroom in Khorezm. A few large state-sponsored mosques were built, but once opened they were little frequented. In Yangibozor city, the district centre (*raytsentr*), there was apparently one such newly built mosque, but no one among the farmers I knew could tell me where it was.

In Khorezm, village mullahs usually lack formal religious education and seldom make public appearances, doing so, for instance, only when requested to say blessings on the occasions of ceremonies. I met the imam of Yangibozor district only once, during the *ekin bayrami*, the sowing celebration held when the cotton seeds are planted, a ceremony introduced under the Soviet Union. The imam's presence had been requested so that he could say the final blessing at the close of the meeting, which was de facto a *pokaz*, or seminar, held by the district governor to give instructions to the inner circle of district dignitaries for coordinating the sowing campaign in 2004. Unlike what Abashin (2006) observed for the Ferghana Valley, as far as I could see in Khorezm, the role of the mullah as a defender of local notions of morality was marginal and did not drive or challenge customary habits in the villages.

It appears, then, that Khorezm has retained fewer traditional social institutions than the religiously and socially more conservative areas of eastern Uzbekistan, as if in Khorezm the rural communities have lost the wealth of their traditional social fabric. One might speculate that the small size of this irrigated oasis, surrounded by deserts, and its remoteness from other centres of the area enabled a more efficient repression of indigenous social institutions during the Soviet years, but I found no hard evidence for a tradition of 'weak normative order' (Geiss 2003: 146) characterising Khorezmian communities. Because there is little hope of finding an ultimate answer to this question, what remains is that the traditional social infrastructure of Khorezmian communities differs in many respects from that of the rest of Uzbekistan.

The Khorezmian Family, Past and Present

It is probably fair to say that in Khorezm the absence of counterbalancing institutions such as the mahalla and the mosque pushed the extended family to the centre of traditional community life, more so than in places where the mahalla has been strong. The ethnographies produced by the members of the so-called Khorezm school corroborate this view. Dating back to the Khorezm Archaeological and Ethnographical Expedition of the Soviet Academy of Sciences, headed by Tolstov in the 1940s and 1950s[50], Sazonova's work on the material culture of the Uzbeks of southern Khorezm (Sazonova 1952) and Snesarev's works on everyday religiosity, demonology, and popular Islam (1974, 1976, 2003) offer valuable information about the way local communities were organised and ruled before and shortly after Soviet modernisation. Their descriptions show particular interest in the extended family. Snesarev (1976: 67) wrote:

> The extended family was of particular interest to me. Such families, which scarcely exist today, included three or more generations and comprised several dozens of members of both genders. All work was done collectively. Members took turns preparing meals in large cauldrons for the whole family. The term 'Bir Kozon', meaning 'a single, collective cauldron', therefore came to describe such families. The head of such a family would be one of the more senior men whose wife would lead all the women's activities. In Khorezm these

---

[50] For a detailed account of Tolstov and his archaeological and ethnographic expedition to Khorezm, see Germanov (2002). The ethnographic section of the expedition was directed by Gleb Snesarev.

extended families typically lived in an estate called a 'howli' [*hovli*], a fortification with towers, high mud walls, and massive doors, which would block entrance to the court. The patriarchal extended families were the repository of traditions, oral history, legends, and popular beliefs.[51]

Snesarev's descriptions offer a snapshot of a time of transition in Khorezmian communities, a moment when rural people's memories of the pre–World War II traumas of forced collectivisation, 'de-kulakisation', anti-religious policies, and bitterly suppressed local reactions against these must have been still vivid. The Soviet program of social engineering that, according to orthodox socialist doctrine, would raise Khorezm 'from feudalism to socialism, bypassing the intermediate state of capitalism', was only beginning. Change had not yet come as decisively as it would some years later, with the mechanisation of agriculture and the increasing modernisation of society. When Snesarev visited Khorezm, extended families were still common: 'A tendency to live in undivided families is observed in Khiva elats even in our time. For the most part, sons separate only after the father's death. While alive, he is the head of the family and has control of all the income of its members working in the kolkhoz. In any case, a daughter-in-law living in an undivided family is dependent in a material sense on its head' (Snesarev 1974: 229).

Soviet ethnographers dealing with the traditional Central Asian extended family emphasised its resistance to change.[52] In today's families, continuities with and resemblances to past practices remain strong in domestic life in rural Khorezm. In the areas in which I conducted my research, the lives of families reflect what Kandiyoti (1998: 563) called the concept of the rural household. Although no longer as large as they once were, rural families are still extended families including several generations (often three, sometimes more) who live in a joint household and form a unit of consumption, distribution, and production organised around the figure of the patriarch, to whom family members show respect (*hurmat*) and obedience.

The extended family is composed of the patriarch, his wife, their unmarried daughters and sons, and married sons with their wives and children. In terms of status and authority, family members are ranked by decreasing age and gender. Through marriage, daughters change their family affiliation and therefore are not counted as members of their birth family in the same way sons are. This is reflected in kinship terminology, which differs for

---

[51] Author's translation from the German.

[52] For a detailed ethnographic description of the extended (rural) family, see Bikzhanova, Zadykhina, and Sukhareva (1974). Krader's account (1963) was based on Soviet ethnographic sources.

maternal and paternal kin. Rules of kinship and of hurmat are highly valued. Once, when a friend and farmer from Yangibozor invited me to a banquet at his relatives' house in Urganch, the head of the household, a maternal cousin (*jian*) of my friend, at one point asked him permission to leave his own house. My friend explained that the question illustrated the 'beauty of our customs'. Hurmat, he said, required that maternally related relatives show respect to their in-laws – in this case my friend – and that hosts show respect to their guests.

In traditional families, parents arrange the marriages of their sons and daughters. The sons of the head of household move out of the house, with their wives and children, only long after marriage, when the family, in a common effort, has provided them with a new house in a nearby location. Although for married sons it is desirable to achieve their own courtyard house (*hovli*) quickly and move out of the parental house, this can take long time, for economic reasons. Traditionally, it is among the duties of the head of household to provide houses for his sons, but in recent times of economic difficulty, the rules have become flexible and can vary a great deal. In Urganch I knew several sons who, out of need or desire, had made the jump to independent life without their father's help.

In rural areas, establishing a separate household entitles the new family to apply to the local authorities for a plot of 0.13 hectare of irrigated land, called a *qo'shimcha tomorqa*. This plot, together with cattle and fruits and vegetables from the garden around the house, represents the economic basis of the average rural family. Even when sons have moved out of the parental household, the links between them remain strong until the death of the patriarch, who, while living, maintains decisional authority in all important matters concerning his sons. Whereas elder sons progressively move out of the natal household, the youngest son and his family remain with the parents, and he inherits the parental house.

Although Snesarev's descriptions still fit most rural families in Khorezm, an important difference is that nowadays the average number of household members has shrunk considerably. According to Sazonova (1952), another member of Tolstov's expedition, in the late nineteenth and early twentieth centuries the patriarchal extended family, which her interviewees called by the terms *ailya* (Uzb. *oila*, family), *keta-ailiya*, and *kabiliya*, typically comprised 45 to 60 family members living together, unified and ruled by one family leader, the *buva* (literally, 'grandfather'). One of my consultants, a man named Allambergan, a peasant and member of the O'yrat neighbourhood committee (*elatkom*) in Yangibozor, told me that in the past, large courtyard houses hosted extended families in which most sons and their wives and children shared a roof with the family head. Families easily reached 20 persons, and often more. Allambergan was referring to the time of his childhood, the same period when Snesarev visited Khorezm. Krader (1963: 144), on the other hand, drawing on Zadykhina (1952) and Kislyakov (1954: 162), wrote that 'the revolution has changed the family structure' in that 'the extended family form has tended to change in the direction of conjugal component units'.

Extended families today, although still spanning two or more generations, are much smaller than those Sazonova and Allambergan described. According to my observations, average rural households in Khorezm had six to seven members living in a hovli or an *uy* (house).[53] Allambergan explained to me that today, although households live separately, family networks continue to play crucial roles in the material lives and livelihoods of villagers:

> Allambergan: This *posolka* [village] here, for instance, is small; they are all related by kin.[54] So if the one household gets a bad harvest, or if it gets no water for irrigation, in any case they will share the harvest of their *tomorqa* [private plot] between households in an equal way.
>
> TT: If in Bo'ston [a former sovkhoz] there are 1,500 people and maybe 400 households, how many such family groups are there? I mean, those who would share their harvests.
>
> Allambergan: Many. If you count the whole *zona* [a sector of the former kolkhoz] of O'yrat village, for instance, there are 10 to 12 different *avlod*s [groups of agnatic kin], and they are not related. They will not share their harvests except among themselves.

Terms that I often heard used to refer to extended families were *avlod* and *urug'-avlod*, meaning a group of brothers linked by agnatic descent, their families, and, if still alive, the patriarch. The explanation I received for why mothers' and sisters' lines were not considered equal members of the extended family was that 'from the father's side it is all the same blood. The child of your *uka* [younger brother] is also your *uka* [*Ota tomondan bir qon, ukaning bolasi ham uka*]'. The avlod is still a fundamental unit in the organisation of the agricultural world after Uzbekistan's land reforms.

---

[53] The work of Djanibekov (2008a, 2008b: 46), who conducted an extensive household survey in Khorezm, confirms this estimate.

[54] *Posolka* is Russian word meaning a settlement in a kolkhoz. In Yangibozor it is often used as a synonym for village (*qishloq*).

Gender Relations

For Sazonova, prior to the Revolution of 1917, rigidity, conservatism, and 'subordination of the will of the family members' to the buva characterised relations within the Khorezmian family. Among the undesirable practices of the traditional family, she mentioned above all the 'humiliation of human dignity, especially of the youngest and of the women' (Sazonova 1952: 309). Without taking up Sazonova's ideologically coloured vocabulary, one can observe that such characteristics, along with a gender-specific division of labour, still partly live on in today's family relations. For instance, in Khorezm it is still considered impolite for a bride (*kelin*) to speak directly to the senior members of her groom's family during the first years after she moves into her new household. More generally, gender segregation has remained strong, and in rural areas it pertains to the domestic as well as to the public sphere.

The main activity of extended families in Khorezm just before the Soviet period was agriculture, and tasks followed the lines of gender (Sazonova 1952: 301–302). Krader (1963: 145, 162) wrote that among what he called peasant families ('*dihkan*') in Central Asia in the pre-revolutionary period, 'silk raising, dairy production, gardening are women's tasks ... , the men are herdsmen, canal diggers and cleaners, house-builders. Cotton and grain raising are joint tasks among the farming peoples'. This gendered division of labour is reflected in the writings of Sazonova and Snesarev for pre-revolutionary Khorezm. The Soviet state, although it greatly improved the condition of women, did not fully achieve its proclaimed goal of emancipation. According to Kamp (2006: 230), 'the Soviet state did not succeed in transforming Uzbek gender ideals, nor did the state even try; but it did make huge changes to the way Uzbek women lived their lives economically and socially, and some of these changes – ending seclusion, providing universal education, making room for women in public, and providing basic health care – should be recognized as positive'.

Although advancements were made in rural Khorezm, they were less significant than in urban and in less peripheral milieus. Snesarev (1974: 226–227) described the working conditions of women in rural Khorezm in the 1950s:

> Women, at least in those localities where field studies were conducted, are poorly involved in social life, in spite of the fact that their laboring success is well known, and that in the cotton fields they are the main labor force. Women here have failed almost totally to be advanced to supervisory work in kolkhozy. ... However, they themselves refuse this, and also encounter obstacles in this regard

placed by representatives of the male sector of the population: both relatives – fathers, husbands, brothers – and unrelated persons.

Interestingly, unlike in Krader's description, in which burdensome cotton picking is mentioned as a 'joint task', in Snesarev's account it is a task mainly for women. In this respect the Soviet emancipation campaign apparently resulted merely in a broadening of women's labour activities, without the counterpart of their achieving significant social advancement.

The gendered division of labour that Snesarev described has been maintained until today. In the rural areas of Khorezm, technical jobs (tractor driver, engineer), management positions, and leading positions in the administrative staff and in agricultural decision-making are largely the domain of men, and women are seldom represented in them. Rural unemployment, however, has affected the traditional distribution of roles between the genders, and the division of labour in rural families is developing new forms. As many men leave the villages for temporary labour migration, women see their share of work in the fields increasing. They sometimes take over activities traditionally done by men, such as looking after the rice paddies on the small, subsidiary household plots (*qo'shimcha tomorqa*). Kandiyoti (1998: 566) has drawn attention to the 'feminization of poverty' that this process also entails.

Land Tenure and Rural Families during the 'Feudal' Period

Little is known about Khorezmian rural communities during the khanate and tsarist periods. Information is superficial, at times contradictory, and written mainly on the basis of Russian sources that reflect the perspectives of the former colonisers, with all their biases and misunderstandings.[55] In particular, landholding patterns and rural taxation in the pre-revolutionary khanate have so far been little studied. Changing legal cultures and shifting states of property add to the complexity of the issue (Sartori 2010).

However, a long tradition of land regulation and land ownership existed well before the tsarist conquest (McChesney 1996: 54), and according to Yuldoshev (1959), the use of land and water resources in the khanate of Khiva was complexly regulated. Taxation was based on an elaborate system of land rent (*ijara*) that distinguished between categories of rent-ownership on the basis of the size of the rented land estate. Yuldoshev distinguished three types of ijara: *avlo* (more than 10 *tanob*s of rented land), *avsat* (5–10

---

[55] For these reasons, according to Paul (2010: 126), 'it is next to impossible to understand pre-colonial landholding from the paper trail left by the colonizers; we have to look at the documentation in native languages instead' – a task yet to be fulfilled.

tanobs), and *adno* (less than 5 tanobs).[56] According to Guliyamov (1959), in the second half of the nineteenth century, before Khorezm became a tsarist protectorate, Khorezmian peasants were burdened by extremely high taxation. He named 25 kinds of tribute and duties that peasants were supposed to meet, which supposedly left them with little means to survive (Guliyamov 1959: 246–247).

Neither Yuldoshev nor Guliyamov looked much deeper into the way the land tenure system and its power relations actually worked. Soviet writers referred to the khanate as the period of feudalism, in which a backward elite of bey landowners was juxtaposed against the exploited rural masses. In an attempt to reflect class relations in rural Khorezm, Sazonova wrote about two extended families, the Po'staklar (the Po'staks) and the Maxsumlar (the Maxsums), as representatives of the supposedly antagonistic classes of rich and poor peasants, at a time when extended families were composed of related households still living jointly, 'under one roof', in a large house. She described the two families as follows:[57]

> A poor family – the Po'staklar
> The family Po'stak owned 12 tanobs of agricultural land. This was far too little for a family with 40–50 persons. The land was not enough to ensure the needs of the family, which lived in poverty. For a long time the Po'staklar could not afford to have their own house so that they had to live in a hut made of mud. Three water wheels, situated in different parts of their land plot, were used for irrigation. The family kept two camels for the maintenance of the water wheels and for irrigation. The land was ploughed and worked on with the help of two oxen. Moreover, the family owned two draught horses. The adult and adolescent male family members did all agricultural work. Women helped out only during the harvest, when a large labour force was needed at once. Their agricultural work consisted in the harvesting of wheat and cotton and in preparing the hay and clover reserves. Moreover, women were responsible for the entire domestic work.

---

[56] The tanob was the unit of land measurement used in the nineteenth century. One Khiva tanob equalled approximately 2.5 hectares. Matley (1967: 278) defined a *tänab* as 'a measure of land that varied in size from two-fifths of an acre to one and a quarter acres depending on the locality; in Khiva, for example, a *tänab* was slightly larger than one acre'.
[57] Both paragraphs were translated from the Russian by Alwin Becker.

A wealthy family – the Maxsumlar[58]

The Maxsum family, according to the information of the elder Ibadullo Yahiyev, who presumably gave the size of the land as smaller than it was in reality, owned 36 tanobs of *mulk* land (privately owned land). Before the revolution the family was among the richest bey families in the area. Within the family, the division of labour was strictly regulated. In all agricultural matters the first assistant of Zikri, the buva, was his brother Yahyo, whose age was closest to that of Zikri buva. The family land was actually worked by three men. The eldest (the *koshka band*) was Rajab (the second son of Abdukarim, who was born third among the brothers, after Zikri and Yahyo), who was responsible for the draught oxen. Rajab was considered to be very experienced in all questions pertaining to agriculture and to the working of land. His regular assistants were his brother Jabbar and Qo'chqor, the son of Aziz, an uncle on his father's side. The *arbakesh* [Uzb. *arabakash*, cart driver] were the aforementioned Aziz, Ziyo, and Seul. The last two were the sons of the second brother, Yahyo.

Sazonova's ethnography highlighted, especially in the case of the 'wealthy' Maxsum family, the way kinship relations translated directly into the labour relations of farming. Moreover, she paid careful attention to all manifestations of social differences that, in the pre-revolutionary context, could be imputed to an endogenous form of class divide. What emerges from her data, however, is the difficulty of translating the attributes of 'bourgeois' and 'proletarian' into the peasant world of Khorezm, despite the fact that Khorezmian rural society was indeed stratified. Her comparison of two cases of extended families representing wealthy (bey) and poor (*batrak*) agriculturalists, upon whom land and water reforms, first, and collectivisation, later, acted with the declared aim of mitigating inequalities, brings to the surface how the social differences of the peasant world – being more a matter of degree than of kind – were inadequately understood in terms of antagonistic classes.

In Sazonova's descriptions, some insights into the working of the traditional Khorezmian *oila,* or extended family, appear. The buva, the family patriarch, ruled the family with unrestricted authority. After his death, the oldest relatives decided who was to be appointed as the family's new buva. If the oldest person in the family had too little authority, then the council of family elders appointed another person among the eldest – the one who was

---

[58] The name Maxsum reveals the family's prestigious origin. According to Abashin (2006: 272), *maxsum* is a title for those 'recognized as belonging to families educated in religion and recognised as mullahs'.

most active and suitable to rule the family. The slight age difference between the two candidates did not matter (Sazonova 1952: 303).

The death of the buva had no repercussions on the property and other assets of the family. According to customary practice, possessions were not distributed among the family members; a new person was simply appointed at the top of the family structure. No one in the family owned money or property on his own. So, for instance, in the bey family Maxsumlar, it was Zikri buva himself who went to the bazaar to buy everything the family needed from it, although for most staples the family was self-sufficient. Dishes were purchased from the potter. The shoemaker and the barber moved from village to village and served people in place. At harvest time they came and asked for payment for their cumulative yearly work, for which they received 20 kilograms of wheat from each family. According to Sazonova's interviewees, no family member had a source of income separate from that of the rest of the family. If, for example, a family received bride price (*qalin*) for a daughter at her wedding, it was treated not as the property of the young woman's father or mother but as income for the whole family (Sazonova 1952: 305).[59]

Ethnographers of the Khorezm school mentioned evidence for the evolution of capitalist relations in the khanate of Khiva, suggesting that a deep transformation of the extended family was under way even before sovietisation. According to Sazonova, the decay of the extended family took place rather quickly: it was 'eroded from within' through the introduction of private property in the means of production. Under these circumstances, the death of the family head (buva) often became the reason for the redistribution of family property, so that the introduction of private ownership, together the different positions of older and younger family members in regard to property, put the extended family under pressure well before the Bolsheviks came to power. Nevertheless, Sazonova wrote, there were reasons to believe that the khanate government, essentially for tax reasons, purposely hindered and thus artificially prolonged the decay of the extended family. For purposes of taxation, each extended family was treated as a single unit, which acted as a disincentive for the splitting of families while making the work of the tax collectors easier (Sazonova 1952: 306).

Sazonova's ethnographic material, drawn from interviews about extended families during the time before the Revolution of 1917, portrayed pre-socialist family relations in Khorezm. Rural families there, although equipped differently in terms of belongings and land, shared the same type of family structure and lifestyle. This circumstance was manifested in the

---

[59] This paragraph is based on a translation from the Russian by Alwin Becker.

difficulty Sazonova had in adequately framing the forms of exploitation existing in rural society in terms of class relations. Rather than refer to exploitation in the relations between extended families of different statuses, Soviet ethnographers emphasized forms of exploitation within the patriarchal family.

From *Yakka Joy* to *Posolka*

Khorezmian houses were built according to a precise plan. According to Sazonova, the peculiarity of the traditional Khorezmian hovli was that the whole building complex, the size of which could range from 300 to more than 2,000 square metres, was built under a single roof. Thus it differed from traditional dwellings in other parts of Uzbekistan, where the rooms of the house were usually built around an internal courtyard. She also affirmed that traditionally all Khorezmian houses were divided into a male area, situated close to the entrance of the house, and a female area, in the middle of the building (Sazonova 1952: 282). Moreover, she wrote that the 'large patriarchal family', a 'remnant of feudal society', lived according to the principle of economic self-sufficiency and had all the characteristics of an 'isolated cell of society' (Sazonova 1952: 289).

Within an elat, or neighbourhood, the construction of a new house was the duty par excellence in the custom of *hashar,* the voluntary work in which people engaged to accomplish tasks for the common good of the community.[60] The elders of the elat would decide when the construction of a large farmhouse should take place, and the whole elat would take part (Snesarev 1974: 228). The participating workers were paid only with the food they received during the workdays, but the family that asked for help was implicitly obligated to reciprocate by helping others in the future. In Sazonova's descriptions of traditional house-building in Khorezmian communities, usually 15 to 20 male workers were involved in well-organised construction. Two workers mixed clay for *paxsa,* the clay blocks normally used for such dwellings, while three brought more clay to the construction site on a high, two-wheeled cart called an *araba*. These were the subsidiary workers. Teams of eight persons carried out the main work, which consisted of placing the layers of clay blocks. Two men were charged with moulding the delivered clay into blocks. Another aligned and adjusted the blocks in place

---

[60] The definition of *hashar* is fluid, and calls for hashar nowadays are much misused. For instance, during my fieldwork in Khorezm a teacher referred to the mandatory cotton harvest campaign as hashar that the people of the *rayon* were doing for the state. In the past, this use of the term was adopted to justify the compulsory mobilisation of workers for large construction projects such as digging irrigation channels. In Yangibozor I heard people distinguish between state-mandated hashar and 'little' hashar for the community or the family.

as the wall rose. If the house owner's means permitted, another team of eight, working in the same manner, would be called in. In addition to relatives, cohorts of male friends and neighbours of the same age (called *jo'ralar*) joined the work. Every *jo'ra* contributed by working from two to three days at the building site. The construction was steered by a senior supervisor, an experienced older member of the community (Sazonova 1952: 290).

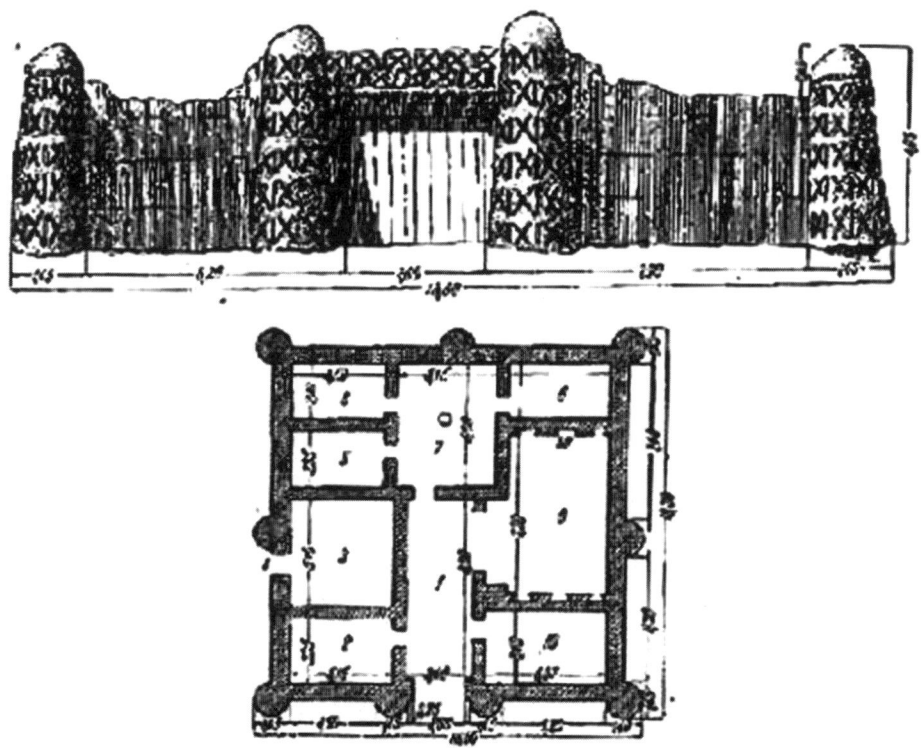

Plate 4. The house of Yoqubboy Jumaniyozov, an example of a traditional Khorezmian house. Reproduced from Sazonova (1952: 283).

According to my data, the last large farmhouses of the traditional type in Yangibozor, built to host extended families, were erected in the 1950s. After that time, a new kind of settlement, built by the kolkhozes on centrally allocated land, gradually replaced the old style of housing. The state's declared aim for the new building style was to facilitate the introduction of modern utilities – electricity and, later, natural gas – in rural areas, but the central planners also had a hidden agenda, which was to strengthen the nuclear family to the detriment of the traditional extended family.

On one occasion, Allambergan brought me to visit the old house of his parents, calling it *konya joy* (old place) and *yakka joy* (single, individual place). He explained that this was the kind of house built before the Russian model of the *posolka,* a more regular settlement with houses built close to one another along the street, became predominant. Built in the 1950s, the house had been abandoned for decades and was now being used as a storage facility. According to Allambergan, the house was among the last in Yangibozor to be built in the old style. The building was a much more modest and modern version of an old Khorezmian 'bey' house. Yet in the photograph I took of it, a pattern of incised vertical lines is visible on the old clay blocks, reminiscent of the old ways of adorning house façades. Because such houses were built without stone foundations and were exposed to the infiltration of groundwater, they had lives of approximately 30 to 40 years, after which they were demolished and rebuilt.

Plate 5. An old house in O'yrat village, part of the Bo'ston MTP, Yangibozor district.

It was unclear to me whether the traditional fortified farm court (plate 5) that people in Yangibozor refer to as a *xutor* was equivalent to the Khivan *avva* to which Snesarev (1974) referred, although this seems likely to have been the case:[61]

> In addition to the principle of settlement by individual 'avva' villages included within the limits of a given kolkhoz, which is characteristic of these localities, the population is divided into specific groups, bearing the name of an 'elat' (community). Each group has its name and lives within the boundaries of its own village, the name of which coincides with the name of the elat. The elat unites a number of families, averaging from 20 to 40. Membership in a specific elat is still distinctly recognized by the population; the names of the elats are used in ordinary conversation. Especially characteristically, however, they disappear altogether when the conversation refers to productive activity; in these cases only the names of brigades and teams figure (Snesarev 1974: 228).

Ethnographers such as Sazonova and Snesarev were fully aware of the importance of settlements in the transition from the 'traditional' to the 'modern' mode of organising the collectives. They addressed the type of housing as an indicator of the degree of success of modernisation policies, and they themselves recommended housing to party officials as a target for these policies. According to Snesarev (1974: 282):

> The Khorezmian farmhouse maintained its general character as the settlement of the patriarchal extended family until the great socialist October Revolution. This type of diffuse settlement can be found in Khorezm up to today. But nowadays, as we see the beginnings of construction of the socialist kolkhoz settlements, the general character of the settlements is changing before our eyes, in the direction of increasing compactness and determinedness in regard to the location of the single houses.

The chief characteristic of the new mode of housing, which transformed the look of the Khorezmian countryside, was the density of settlements, in which new houses were built compactly within a clearly defined area. Originally, houses in Khorezmian settlements were built along irrigation channels that carried water to fields and orchards, so they were scattered loosely over the landscape. In a description of Khiva, Sazonova (1952: 277)

---

[61] In Russia, *khutor*s were 'the individual enclosed farms separate from villages whose establishment had been encouraged by the pre-Revolutionary Stolypin reforms' (Fitzpatrick 1994: 163). They were common in western Russia, Ukraine, and Siberia. According to Roy (1997a: 145), Sharof Rashidov in 1964 announced to the plenum of the Communist Party of Uzbekistan the launching of a battle to eliminate the khutor system.

wrote that people settled along channels and water ditches, with small gardens in the backyards and at the sides of their houses. Trees planted along the channels and ditches supplied the inhabitants with timber for building houses, boats, and bridges and for making work tools and other utensils. The irrigated fields were situated around the houses or in their proximity.

The new houses were designed for smaller families, with three to four rooms. They were planned by professional technicians and built with the help of the kolkhoz. Doors and the inlaid columns of the '*aywan*', or veranda, were often taken from the old houses and integrated into the new ones. For Sazonova, the most important change was that the houses, now with windows in their walls, acquired 'openness' and no longer resembled the 'closed fortresses' of the past. Inside the new houses one could find furniture 'typical of the urban lifestyle, such as sofas, wall clocks, LP-players and the like', suggesting that rural Khorezmians accepted and welcomed their new lifestyle (Sazonova 1952: 316–317). Snesarev, on the other hand, wrote about signs of recalcitrance among members of traditional society to the new settlement pattern: 'The prevalence in our day, although in a residual form, of these territorial-kinship groups introduces certain complications in the social life of kolkhozy. For example, the designing of new-type settlements encounters difficulties due to the fact that, first, elats do not wish to abandon inhabited places, and, secondly, the existence of friction between elats and, in contrast, the kinship ties among groups, complicate the organization of new settlements' (Snesarev 1974: 229).

With the shift from yakka joy to posolka, and with the enlargement and better equipping of the kolkhozes, the Khorezmian landscape was radically altered, and only a few landmarks of the past have been preserved. Yet 'territorial-kinship groups' have survived and play a role even today. In Khorezm I learned that people's relationship to their extended family was a source of pride, even if the family's name denoted a low-status origin. For example, a man called Zarif-qul (literally, 'Zarif-slave'), the rais of a kolkhoz with which I worked in Yangibozor, proudly explained to me that his name revealed his family as descendants of slaves, *qul* being a nickname (*laqab*) that was common in his village of origin.[62] Many names of *fermer* enterprises also reflected the owners' affiliation with particular territorial-kinship groups. Among the fermers with whom I worked closely, I found, for instance, that in Madaniyat the fermer enterprise called 'Rahimboy dev' was named after the buva, Rahim, of the extended family known as the

---

[62] Kehl-Bodrogi (2008: 53–54) gives a different reading of the connotation of *qul* in Khorezm. In Yangibozor I learned that although typically a laqab referred to a physical characteristic or a profession, the term *qul*, which denotes descent from slaves, was also used as nickname.

Devs, one son of which owned the farm in 2004.[63] One of the larger rice-growing fermer enterprises with which I worked was called 'Urin-qoq'. When I asked about the origin of the name, the farm owner told me he had selected it to honour the memory of his father, Urin buva, who had been a well-known brigadier of the former rice-growing sovkhoz 'Rossiya' and whom people knew by the nickname 'qoq'.

Socialist Modernisation in Khorezm

As was the case all over the Soviet Union, so in Khorezm the strongest impulses for change were transmitted to rural society through the early land and water reforms (1925–1926) and through collectivisation (1928–1932). As Teichmann (2007) put it, 'peasants and local officials had to look for ways to reconcile large-scale cotton production with their own survival needs'. Because of the scarcity of sources, reconstructing the concrete forms this 'reconciliation' took in Khorezm is difficult. In some of my interviews, people recalled the early Soviet years as a time of individual or 'private' farming (*yakka xo'jalik*), characterised by famine (*ocharchilik*). I found no sources specific to Khorezm or Yangibozor regarding the way local communities and authorities interacted in the early collectivising of agriculture, nor literature about the local implementation of the early reforms. Therefore, my remarks about the dynamics of collectivisation in Khorezm must be couched largely in general terms.[64]

In one of the few descriptions of the early Soviet reform period in Khorezm, Battersby (1967: 23–24) wrote:

> Between 1925–1928 land and water reforms were supported by the Communist party, and while beylandlord land titles were abolished, the land thus previously held was given over to peasants or kolkhoz(es) (collective farm(s)). By 1948 the mechanization of the kolkhozes was accomplished, but the peasants continued to use the kunde-, omach-, and sokha-ploughs, and the mala-levelers, and other such primitive tools, to cultivate their personal land 'as in previous years'. Since the Communist take-over, energy has been expended in draining swamps and in repairing old and building new irrigation systems. Horticulture and tree planting have been encouraged. Fruit

---

[63] According to Snesarev (1976: 29ff.), in Khorezm the *dev* were a sub-category of spirits ('jinn') renowned for their strength and trickiness. The nickname *dev* denoted the family's supposed physical strength.

[64] A body of literature does exist, however, on the first Soviet land and water reforms and subsequent collectivisation in the Uzbek SSR. See, for instance, Thurman (1999: 14ff.), Khan and Ghai (1979: 37ff.), Matley (1967: 289ff.), and Teichmann (2007).

trees and others of economic importance, such as the poplar and mulberry, the latter for sericulture, have been planted.

Such a description obscures the hardships and sacrifices the local population must have faced with the expropriation of their land and belongings and the compulsory handover of foodstuffs, harvest quotas, and farm assets (Fitzpatrick 1994), and it highlights results achieved by the collective farms only years later. Interestingly, it also shows how Soviet modernisation was a gradual endeavour in which old and new modes of production long coexisted in the communities, and with them, old and new modes of social organisation.

The protocol of the annual assembly of the Kuybyshev kolkhoz, one of 10 kolkhozes that would later merge into the Bog'olon kolkhoz in Yangibozor district[65], reported that in 1938 the kolkhoz had a population of 225, of which 112 persons were more than 16 years old, 7 were between 12 and 16, and 8 were outside the kolkhoz for study, military service, other official duties, or reasons unspecified in the document (possibly they had been jailed or deported, as district staff suggested). According to the document, the kolkhoz had 118 members, 86 of whom attended the meeting. The number of households is given as 55, but the number of houses and buildings in the kolkhoz is not given. Many residents at the time must have been living 'under one roof' and not in separate houses.

From the protocol it appears that besides the kolkhoz chairmanship, there were only two higher-management positions: those of bookkeeper and head of the 'revision commission', positions held by a father and son, respectively. The document states that the kolkhoz comprised 135.5 hectares of irrigated agricultural land, of which 102 hectares were sown in cotton, 3 in rice, 1 in pulses, 3 in sorghum, 0.5 in vegetables, 2 in melons, 7 in seeds of an unspecified kind, and '7 plus 10' hectares (*sic*) in fodder crops. In 1938 the kolkhoz had neither tractors nor cars, but the listed livestock included 2 cows, 33 oxen, 4 Karakul sheep, 4 camels, and 29 draft horses, as well as 34 carts (*aravalar*), 5 horse-driven ploughs (*atli pluglar*), and other traditional agricultural tools the meaning of whose names I could not identify.

---

[65] This document is in the Yangibozor district branch of the Khorezm state archive. The protocol is dated 17 January 1939 and was signed by the rais of the kolkhoz. According to the Bog'olon village booklet (*Xorazm viloyati Yangibozor tumani Bog'olon qishlog'i pasporti*, 2004) the Bog'olon village administration was established in 1928 and at first comprised the kolkhozes Namuna, G'airat, Qizilqalam, Komsomol, Kalinin, Qiziloy, Qisilko'shni, Pioner, Faoliyat, and Maydamillat, but the names of kolkhozes changed frequently and kolkhozes gradually merged. Finally, in 1959 Bog'olon 1 and Bog'olon 2 merged into a single kolkhoz, ending the consolidation phase.

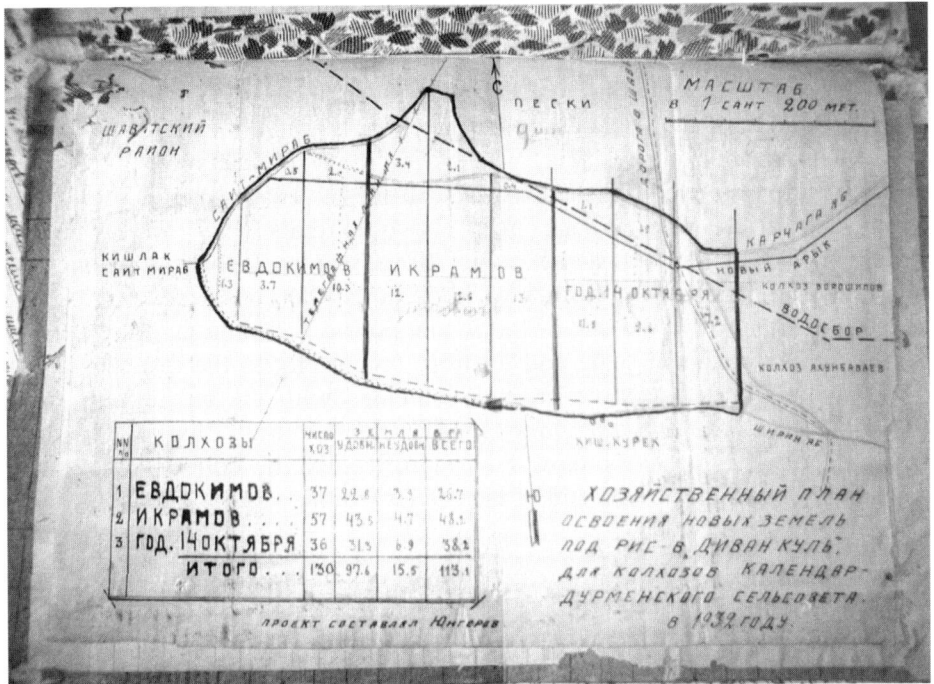

Map 2. Qalandardo'rman sel'soviet in 1932. The inscription at bottom right says, 'Economic plan for the cultivation of new land foreseen for the growing of rice in Divankul', for the kolkhozes of the sel'soviet Kalendar-Durmen in the year 1932'. The table refers to the populations and the quality of land of the three kolkhozes, Yevdokimov, Ikramov, and 'Fourteenth Anniversary of the October Revolution', that made up the sel'soviet. The three together counted 130 households and 113.1 hectares of agricultural land, of which 97.6 hectares were of good quality and 15.5 of low quality. Map scale is 1:20,000 (1 cm = 200 m). Source: Yangibozor district archive.

By comparison, in 2004 Bog'olon village extended over an area of 4,500 hectares.[66] Divided into four elats, it counted 2,222 households and 9,905 inhabitants but probably housed even more.[67] In the more than six decades separating these two sets of statistics, the Khorezmian countryside had been radically changed. A perhaps even more powerful illustration of the deep transformation of settlement and agricultural patterns caused by Soviet modernisation drive can be obtained by comparing a map of Qalandardo'rman sel'soviet (map 2) and the cadastre map of the *shirkat* Bo'ston

---

[66] Since 1959, Bog'olon sel'soviet has been co-extensive with the collective enterprise, a kolkhoz named after Kuybyshev until 1992, afterwards a shirkat, and since 2003 the *motor traktor parki* (MTP) Bog'olon.
[67] Source: *Xorazm viloyati Yangibozor tumani Bog'olon qishlog'i pasporti*, 2004.

(map 3). They document, respectively, the situation at the beginning and at the end of Soviet modernisation in Yangibozor.

The older map, dating to 1932, shows the rice production plan for the three kolkhozes of the Qalandardo'rman sel'soviet, in what is now Yangibozor district, in their early stages, well before their amalgamation into larger units, which was accomplished in the district by 1962. Map 3 shows the decollectivisation plan of one of the two shirkats (the former shirkat Bo'ston) of Qalandardo'rman sel'soviet 70 years later. Insofar as it concerns the territorial setup of the former sovkhoz, it can be considered representative of the situation at the end of the Soviet period.[68] Although the two maps do not fully overlap in scale and territory (according to the director of the Yangibozor District Archive, map 2 coincides with the territory of what today is the *motor traktor parki* [MTP] neighbouring that of Bo'ston, Hamza), they nevertheless show how significantly the collectives evolved from their beginnings to the end of the Soviet Union.

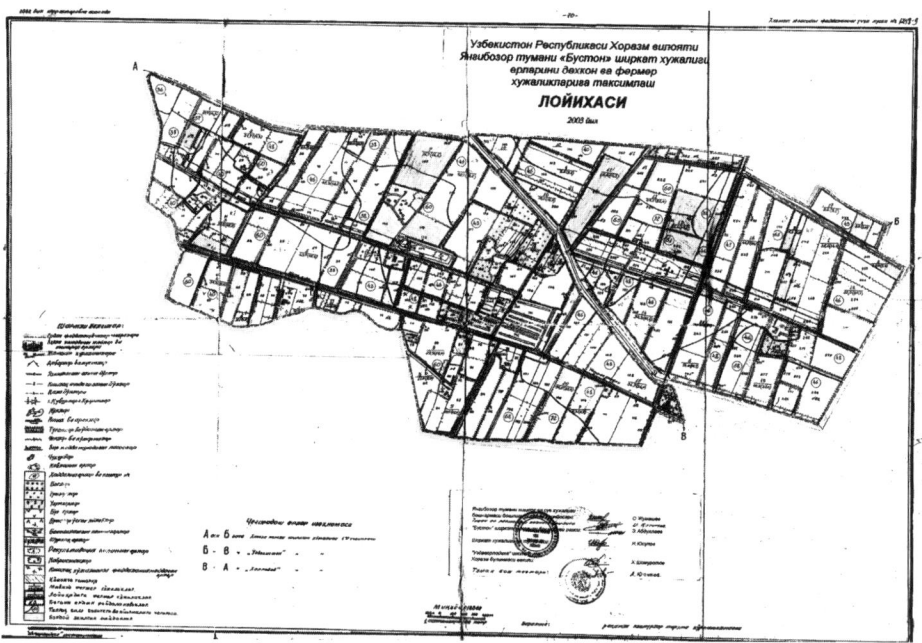

Map 3. Bo'ston shirkat cadastre map, 2003. The inscription reads, 'Plan of the repartition of the land of the shirkat Buston [sic], Yangibozor tuman, Xorazm viloyat, Republic of Uzbekistan, to the fermer and dehqon enterprises'. Source: Yangibozor District Land Measurement Office.

---

[68] According to the land measurer from whom I obtained this map, it was produced at least 10 years before 2003 and was then updated and amended year by year.

In the older map, the boundaries of the three kolkhozes are sketched in schematically, with heavy lines, and brigade boundaries are drawn with thin lines. Within the brigade areas, the portion planted in rice is noted cumulatively in hectares, and secondary irrigation channels are not taken into account. The map also shows a telegraph line, the names of two villages (*qishloq*s), and one street. North of the sel'soviet territory, the inscription *peski* (sand) testifies that large parts of the territory of today's Yangibozor were neither irrigated nor inhabited at the time. Administrative boundaries follow the courses of the major channels and were made by a simple division along the north-south axis.

This way of partitioning territory contrasts sharply with the situation revealed in the second map. It shows how, as a consequence of full mechanisation, land was both intensively cultivated and densely urbanised by 2003, at least by rural standards, with new roads, new settlements, new waterways, and other infrastructure. The older map depicts a situation in which land was still 'opened by hand' (*qo'lda ochilgan*), as an aged peasant in Madaniyat village put it during an interview, and in which fields, situated between pieces of fallow land, were few and small. By the late twentieth century, virtually all the available surface area had been turned into cultivated or built-on land, all plots (*kontor*s) had become visible in planners' maps, and boundaries were being drawn by tractors and other machines. The comparison suggests how drastic the changes were and the degree to which Soviet modernisation changed the landscape and the environment.

Soviet modernisation increased the available space of the rural world, the scale and intensity of agriculture, and, more generally, the opportunities for life in rural society. This latter characteristic relates not only to the new liberties and opportunities for education, travel, and culture that rural people progressively acquired but also to the sheer size of local society. Whereas the three kolkhozes of Qalandardo'rman sel'soviet counted 130 inhabitants in 1932, the same village administration, formerly the sel'soviet, encompassed no fewer than 6,176 persons in 2004, with negligible permanent migration into or out of the area.[69] However reminiscent of the traditional mode of communal organisation the Khorezmian rural communities may be, they simply are not the same kinds of communities they once were. The qualitative leap in the scale of local community triggered many repercussions in the organisation and structure of such communities.

---

[69] Source: *Xorazm viloyat Yangibozor Tumani Qalandardo'rman qishlog'i pasporti*, 2004 (official statistics of the village administration). According to the employees of the village administration themselves, the statistics underestimate the real figures, which are likely to have been 10–15 per cent higher.

At the local level, a crucial Soviet concern was to eradicate the pre-existing political structure and replace it with a new one. The break-up of extended family units, perceived to be backward if not downright hostile towards the Soviet authorities, was intended to break the ties of the traditional community. In Khorezm, the extended family had long been the crucial institution that locally determined most social and economic relations and that affected the way in which traditional society conceived of agriculture, community, and (patriarchal) authority. Throughout the Soviet period, this institutional landscape was gradually transformed, and extended families adapted to a progressively changing countryside.

As a consequence, a new stratification of power based on the management and control of the kolkhoz was introduced. An important effect of the kolkhoz was that, together with the creation of new infrastructure, it generated new local structures such as brigades and new roles of local leadership. The rais, or kolkhoz chairman, the other executive staff of the kolkhoz, and the *brigadir,* the chief of a brigade, became authoritative rulers of the community. These newly created positions held the real power and superseded the old authorities of the community, whose roles, according to Snesarev (1974: 230), became confined to the ceremonial sphere: 'In prerevolutionary times, "yoshully" (as this group is commonly named), together with the aksakal and mullah, supervised the entire life of the elat: economic, legal, and religious. The most important matters were resolved at meetings of the yoshully. ... At the present time, the functions of the group of the most elderly are limited chiefly to the ceremonial sphere'.

Snesarev's word *yoshully* is *yosh ulli,* which in the Khorezmian dialect means 'old aged', that is, elder or senior. Snesarev used it as the Khorezmian equivalent of *oqsoqol,* 'white beard' or elder, in the mahalla. He noted that in the elats he had visited, there was 'a figure reminiscent of the former aksakal [*oqsoqol*]. In several elats we investigated, the so-called "yoshullysy" (eldest of the elders) is selected from among old and authoritative persons, whom individuals belonging to the elat must unreservedly obey. Thus, in one of the Khiva elats this role was filled by the elder brother of the chairman of the local soviet; in another – by the chairman of the harvest council' (Snesarev 1974: 231). He specified that the 'functions of the yoshullysy consist of overseeing the behaviour of elat members, settling any misunderstandings, etc; however, their activity never goes beyond the boundaries of the elat' (Snesarev 1974: 232).

The term *yoshulli* is a good example of the way pre-existing notions of authority have mingled with new ones, because nowadays in Khorezm the yoshulli is widely understood as the 'boss', in the sense of the person on whom one materially depends and whom one must unreservedly obey. It is not the village elders who have been called this since independence but the heads of the successors to the kolkhozes – the *shirkat*s and *motor traktor parki*s – and the directors of public administrative offices (*tashkilot*) such as banks, tax inspection offices, district statistical offices (Raistats), and the like. In the same fashion, the managers of today's MTPs and farmers' unions are respectfully called *raisbuva,* a term in which the language used to express traditional power relations is adopted into the terminology of Soviet leadership.

For Khorezm and for Yangibozor, the scarcity of well-documented, local-level sources on the crucial events of the Soviet period are obstacles to reconstructing the local evolution of key social relations during sovietisation. This lack of data can be seen as a further sign of the area's peripheral status within the overall framework of Soviet Central Asia. Nevertheless, I found it possible to trace the profile of change in relevant social relations across the history of Soviet Khorezm, relying mainly on the writings produced by researchers of the Khorezm school. Sanctioning the end of the traditional extended family, Soviet modernisation of the Khorezmian countryside reshaped community around collective farms (kolkhozes) by introducing new ways of housing, working, and living. In the process, the collective farms became dominant both as triggers for the economic transformation of agriculture and as vehicles for the social modernisation of community life.

Chapter 3
The Late Kolkhoz Years in Soviet Uzbekistan

Characterised by the abrupt expansion of industrial agriculture and the adaptation of local communities to a rapidly changing environment, the 1960s, 1970s, and 1980s were a period of advanced modernisation in the Soviet countryside. In Uzbekistan during the late socialist period, the form and content of agrarian relations maintained a semblance of continuity with frameworks and practices introduced with collectivisation three decades earlier – hierarchical organisation of collectives, state-directed economic policy, and specialisation in cotton. But the scope of the transformations that occurred, both quantitatively and qualitatively, rendered collective farms in Uzbekistan quite different from what they had been before.

The two cornerstones marking the beginning and end of this late era of Soviet agricultural expansion were the March 1965 plenum of the Central Committee of Communist Party, during which the guidelines of rural development policy for the decades to follow were outlined, and the public prosecutions that gained notoriety in the mid-1980s as the 'cotton scandal', or 'Uzbek affair'. The latter resulted in a pervasive reshuffling of the ruling class of the Uzbek SSR at both the national and local levels, after it was accused of systematic involvement in illegal activities.[70]

In Uzbekistan the guidelines of the 1965 plenum represented the ideal of attainability of the goals of Soviet modernisation, as much as the cotton scandal coincided with the acknowledgement of its moral and economic crisis. It is a widely held assumption that the crisis can be ascribed to Soviet economic policy, which, obsessed with the growth of economic indicators regardless of profitability, sustainability, and quality of output, provided the incentives for the blossoming of a 'parallel' or criminal drift in the conduct of local affairs (Rumer 1989). But in light of the evidence I collected in

---

[70] For synthetic treatments of the Uzbek political events of the 1980s, see Fierman (1991) and Rumer (1989). Gleason (1983, 1986) described the guidelines for the expansion of the cotton economy in Uzbekistan and the politics connected with them.

Yangibozor, this image appears to be over-simplistic and does not acknowledge the proper role of local communities.

Against narrow, economistic views of postsocialist transformation, Lampland (2002: 32) argued that socialism 'was not simply a package of bad economic policies, but a complex social and cultural world in which people lived and worked'. She acknowledged the complexity entailed in the routines, practices, and techniques of socialist production and implicitly invited researchers to pay more attention to the complex cultural world in which economic institutions are always embedded. In Uzbekistan's socialist cotton economy, the *rais,* the chairman of the kolkhoz, was a key figure operating in an environment in which, as in the Hungarian case discussed by Lampland – and no differently from capitalist managers – he had to know how to take risks and whom to trust in order to be successful (Lampland 2002: 49). The rais emerged from Soviet modernisation as the key representative of the local agricultural elites, and as an implementer of centrally directed policies, he influenced the way in which modernisation shaped rural society. In this chapter I focus on the role of the local agricultural elites and their interactions with their social environment. Relying on local accounts and data, I discuss the constraints and ideals that informed the local agricultural setting during the years of late socialism, as well as the opportunities and risks local elites took in an environment of expanding cotton agriculture.

Influential studies dealing with Soviet modernisation put their emphasis mostly on the great human costs of its beginnings (Conquest 1986; Fitzpatrick 1994). With respect to Central Asia, they tend to stress the failures and limits of the Soviet modernisation endeavour (Fierman 1991; Rumer 1989). The Soviets' aim first to disrupt and then to radically change local communities was declared and transparent. According to Thurman (1999), it was not collective farming in itself but rather the way in which collectivisation was carried out and collective farms were managed that led in Uzbekistan to a peculiar climate of coercion that went beyond the characteristic problems and shortcomings of the kolkhozes. There, the role of the command-administrative system was more pronounced than elsewhere: 'In fact, there has never been a collective farm system for cotton cultivation in Uzbekistan in the sense implied by the Model Statute. Rather, there has been a system of compulsory production to meet the plan' (Thurman 1999: 43). As a result, cotton farming in Uzbekistan became a permanent anomaly relative to the rest of the Soviet Union, much as in Scott's understanding of the kolkhoz as the emanation of an 'authoritarian high modernism' (1998: 220ff.) that resulted in the 'capacity of the state to essentially re-enserf rural producers, dismantle their institutions, and impose its will, in the crude sense of appropriation' (Scott 1998: 218).

Against this, research results based on locally contextualised perspectives also emphasise communities' role and agency within the Soviet modernisation project (Gambold Miller and Heady 2003; Humphrey 1983). Rather than disappearing, local institutions adapted to the kolkhoz and influenced the way in which Soviet institutions operated and evolved. As a result, the Uzbek kolkhoz became a 'stable hybrid of Soviet and traditional Muslim institutions and practices' (Koroteyeva and Makarova 1998a: 579). A reconstruction of rural life during the Soviet period must take the development of this coexistence into account.

The Kolkhoz and Sovkhoz in Late Socialist Khorezm

Although legally uniform across the Soviet Union, in Central Asia collective farms had their own peculiarity. There, farming was predominantly related to cotton production, which required intensive labour employment and large-scale irrigation schemes. Along with this, according to Khan and Ghai (1979: 42), the average cotton farm in Uzbekistan and Tajikistan had twice the number of households of the all–Soviet Union average farm. Because larger average households lived on fewer salaries, household incomes in Central Asia were considerably lower than in the rest of the USSR. Relegated to the role of cotton producers, rural people there were less inclined towards urban migration than people in the rest of the Soviet Union, even though intensive mechanisation, starting in the 1950s and 1960s, had rendered a large part of the Central Asian labour force redundant. The informal economy gave rural people the possibility of gaining additional income, which explains their tendency to 'remain attached to the collectives for seasonal employment' and to engage in 'household and private subsidiary activity rather than [move] to non-agricultural jobs or to urban areas' (Patnaik 1996: 50), as was more common elsewhere in the USSR. A further peculiarity of the Central Asian Soviet collective was its characteristic 'native regime of consumption', which distinguished the local population, more concerned with expenditures for community-prescribed, life-cycle religious rituals, from the non-native, European population (Koroteyeva and Makarova 1998a: 581).

Towards the end of the Soviet period, the kolkhoz and the sovkhoz, the large collective and state-owned enterprises, respectively, were the central productive units in the Uzbek rural sector, encompassing virtually all arable land.[71] Kolkhozes, owned collectively by their members, could dis-

---

[71] According to Ilkhamov (1998), relying on figures for 1997 from the State Department for Statistics, Ministry of Macroeconomics and Statistics, Uzbekistan at the end of the Soviet period counted 971 kolkhozes and 1,137 sovkhozes nation-wide.

tribute their production surpluses among the members or sell the surpluses at local markets, whereas sovkhozes, which appeared at a later stage in Soviet agriculture, were financed entirely by the state budget and produced directly on the demand of the state (Ilkhamov 1998: 539).[72] Yet the state determined the production plans of the kolkhozes as well and set the prices of their produce, of which it was the monopolistic buyer. Both the kolkhoz and the sovkhoz depended on the centrally set provision of inputs and assets. They represented the framework of production in the rural areas, with a centralised organisation, common capital stock and infrastructure (tractors, machines, buildings, etc.), and often a range of small enterprises producing manufactured goods. Agricultural inputs such as fuel, fertilisers, and seeds, provided by state factories, were distributed and managed at the kolkhoz and sovkhoz level and fell under their accountancy.

In Khorezm, the average kolkhoz extended over approximately 1,500 hectares and employed a staff of several hundred workers and administrators.[73] The key officials were the chairman (*rais*) and the executive staff: a chief agronomist (*bosh agronom*), a land measurer (*zemlemer* or *yer tuzuvchi*), a chief engineer (*bosh injiner*), a chief accountant (*buxgalter*), an economist (*iqtisodchi*), the head of staff (*otdel kadr*), and the heads of the work units, or brigades, who worked closely with the head of staff.[74] The kolkhoz field workers, *kolkhozniki* in Russian, *kolxozchi* in Uzbek, were organised into brigades, which were assigned to specific plots of land that the brigade worked in common. Brigades were stable work collectives and the main operational units for both kolkhozes and sovkhozes. In addition to the brigades that worked on the main agricultural crops (during the Soviet

---

[72] More detailed information about the differences between kolkhozes and sovkhozes can be found in Dumont (1964), Nacou (1958), and Khan and Ghai (1979). At their introduction, sovkhozes were presented as an organisational advancement of Soviet agriculture relative to kolkhozes. De facto there was little difference between their management structures, although de jure the kolkhoz was owned collectively and as such was a 'bottom-up' farm, as opposed to the state-owned, 'top-down' sovkhoz. Sovkhozes had greater mechanisation (more tractors per unit of land) and slightly higher wages and better welfare services; kolkhozes on average employed more workers per unit of land. Dumont (1964: 140) pointedly stated: 'En zones riches ..., ou à production très rentables, il y a gros intérêt à rester kolkhozien. En zones pauvres, ... le sovkhoz paie beaucoup mieux, la retraite et les assurances sociales y sont fort appréciées'.

[73] In Yangibozor district, kolkhozes and sovkhozes averaged 1,500 hectares of agricultural land. The kolkhoz Madaniyat counted 800 workers in 1985. Similarly, the kolkhoz Xorazm had 1,000 workers for 2,000 hectares.

[74] This information was first given to me by a former land measurer of the kolkhoz Madaniyat. Later, staff members of former collective enterprises confirmed this organisation, sometimes with minor variations. The mixing of Russian and Uzbek terms reflects what has been common use in Yangibozor.

period, mostly cotton), others specialised in growing vegetables, breeding cattle, construction, raising silkworms, and other activities. The average Khorezmian brigade had around 80 to 120 workers directed by a *bosh brigadir* (more often called simply *brigadir*), the head of the brigade, who was accountable to his superiors in the kolkhoz administration for fulfilling the harvest target.[75]

During perestroika, in an effort to enhance efficiency and reduce waste, limitations on livestock owned by kolkhoz households were lifted, and the individual leasing of small plots of land (*arenda*) was partially introduced (Visser 2008: 36). In Khorezm, according to an agronomist I interviewed in Yangibozor, by the early 1980s brigades from selected kolkhozes had already begun to experiment with a more family-oriented mode of organisation – the *pudrat* system, a family-based sharecropping system. Introduced by Gorbachev, the *podryad* rent was established within the framework of the brigades. It enabled single rural families or cooperative groups to rent land and implements from the collective farm for a given period in order to increase their production (Shanin 1989: 11).

Within the kolkhoz, decisions about crops to be cultivated and the share of land to be assigned to each crop were taken by the kolkhoz management. So were all decisions about the assignment and management of kolkhoz resources. Ordinary kolkhoz members had no say in decision-making, which rested solely on the shoulders of the kolkhoz chairman. Although the general assembly of the members was nominally the highest executive organ in the kolkhoz, in practice it held only the ceremonial function of ratifying decisions taken by management. As Gambold Miller and Heady (2003: 264) summarised it: 'Collective farms, kolkhozy, were theoretically owned by their members, whose assembly was supposed to be the ultimate authority. But in practice, authority in both state and collective farms was imposed top-down by the state and party and modified in informal ways by the strategies of ordinary members'. In Khorezm, kolkhoz chairmen, often born and raised in the villages in which they served, were chosen and appointed by the superiors in the Ispolkom, or executive committee (*ispolnitelniy komitet*) of the district, and then confirmed by the assembly. Sovkhoz directors were appointed directly by the district authorities.

---

[75] In Yangibozor, the larger kolkhozes comprised 15 brigades on average, including those who were specialised in work such as construction and cattle-breeding. The average cotton farming brigade in Yangibozor worked 150–200 hectares.

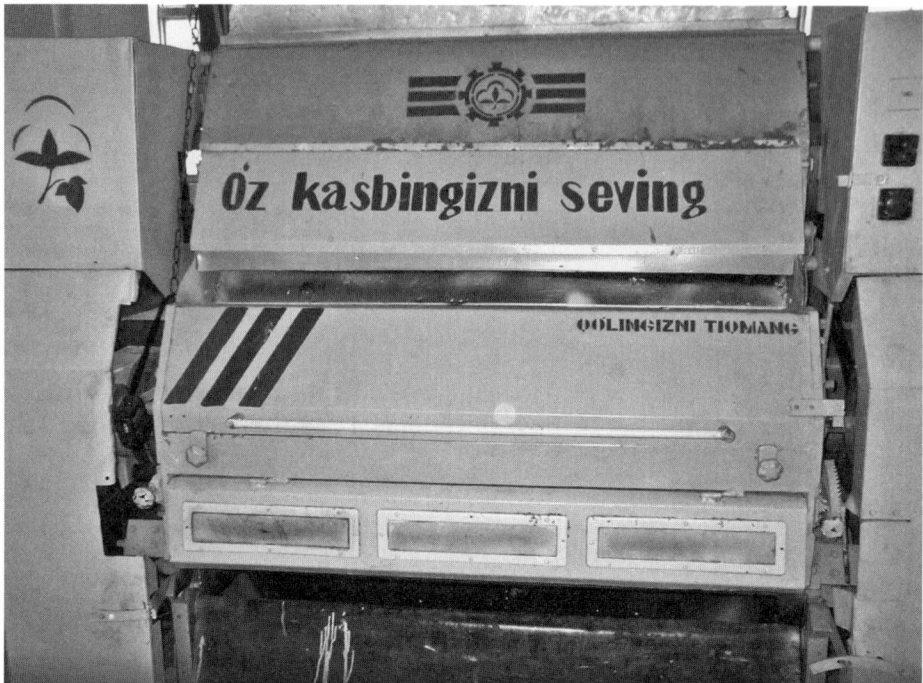

Plate 6. A cotton gin at the processing facility in Yangibozor district. Dating to the late 1980s, it was still in use when photographed in 2004. The inscription says, 'Love your profession' and 'Watch your hands'.

According to the law of the USSR, the state was the owner of all land.[76] Land was granted to the kolkhozes in perennial usufruct for agricultural use.[77] The law prohibited the sub-leasing of collectively held land to individuals for private business. Every rural family, however, received a portion of land for its housing and a small, irrigated, subsidiary household plot on which families could plant subsistence gardens and raise farmyard animals. All kolkhoz workers received a monthly wage from the kolkhoz; sovkhoz workers received their wages directly from the state budget. In many cases the salary was integrated with compensation in kind and harvest surpluses when production targets were surpassed. Produce from household

---

[76] The legislation pertaining to the regulation of agricultural affairs comprised a large number of laws, decrees, and directives by the Central Committee of the Communist Party of the Soviet Union. Above all, the Supreme Soviet, on 13 December 1968, adopted the Fundamental Law on Land of the USSR (see Tulepbayev 1984: 194).

[77] Usufruct is a legal term referring to the right of temporary possession or use of the property of another, in this case the state.

plots formed an important additional source of income, not subject to taxation.

The collective farms also endowed their members with a variety of welfare and educational services and even housing, which in the last decades of the USSR was often provided by the kolkhoz administration. The collective farms of the Soviet Union differed in this respect from those in eastern Europe; in the USSR, 'members received meals, housing, health care, schooling, vacations, pensions, and other benefits directly through their farms, in an arrangement much more comprehensive and generally advantageous than that in Eastern Europe' (Verdery 2004: 142). Unemployment did not officially exist. Every 'child of the kolkhoz' ended up on the kolkhoz payroll, even if many people were provided with work that was useless or superfluous (Patnaik 1996: 52).

In Khorezm, as everywhere in former Turkestan, collectivisation and the expropriation of rich bey landowners levelled the pre-existing socio-economic inequalities. Wage differences within the kolkhoz were small enough that social stratification could not take dramatic forms. At the upper level, kolkhoz notables such as the chairman and some other officials, the brigade chairs, and the other members of the kolkhoz executive committee received wages up to 30 or 40 per cent greater than ordinary workers' salaries (Khan and Ghai 1979). At the lower level, the simple workers enjoyed living standards that during the Brezhnev years (1965–1982) were sufficient to satisfy their material needs and enable them to live decently – as today's rural workers, who face much harsher conditions, sometimes recall with nostalgia. It was not surprising, then, that among the generation born after World War II, I came across first names in Yangibozor such as Marks (Marx), Oktyabr (October), and 'Melis', the acronym for Marx-Engels-Lenin-Stalin, along with Muslim and indigenous names. Names celebrating personalities and events of the Soviet Union were not uncommon even among the less privileged, indicating that the post-war Soviet period was viewed with a large degree of appreciation and legitimacy among the local population.

Still, it would be misleading to portray Soviet rural society as immune to (even strong forms of) social stratification. Socio-economic discrepancies and inequalities simply took forms other than those of capitalist relations. By showing how even 'basic services, which the government claimed to be providing free, could only be obtained in reality by resorting to private, under-the-counter payments', Swain (1992: 224) illustrated how, for Hungary, social stratification became very concrete, with the reproduction of social disadvantages, but in a more indirect and sophisticated way than in capitalist relations. In the Soviet Union, access to non-monetary benefits and to privileged jobs, control over scarce goods and 'manipulable resources' (Humphrey 1998), the power to steer bureaucratic processes, and even the capacity to divert collective goods to the informal, black-market economy were among the factors that created, despite the egalitarian rhetoric, significant economic disparities and social stratification.

Following the campaign of amalgamation of collective farms launched in the early 1950s, collective farms were turned into agro-towns, providing comprehensive services and goods to their inhabitants and aimed at transforming the rural milieu into an industrial one (Nacou 1958: 188). In late socialist Khorezm, every collective farm had a 'house of culture' (*madaniyat uyi*) for concerts and cultural events, kindergartens, schools, libraries, and social and medical assistance services, provided by or supported through the kolkhoz or built on kolkhoz land. In Yangibozor district the collectives were almost the only sources of jobs and housing for rural people. They organised entertainment, recreation, and celebrations of public holidays. Sometimes kolkhoz boundaries overlapped those of pre-existing communities in which Khorezmian people had lived since the time of the khanate. Because of this, and thanks to the all-encompassing services provided to their members, the kolkhozes reinforced the continuity and the stability of these communities and acquired the character of a relatively self-sufficient 'total social institution' (Humphrey 1998; Visser 2008: 184), relatively cut off from the rest of (urban) Soviet society.

Besides being a production unit and a sustainer of rural communities, the kolkhoz was first and foremost a political-administrative cell of the Soviet state, linking the rural world with the chain of the Soviet institutional hierarchy. The position of chairman (*rais*) was one of power, entailing control over many resources and jobs. The preferences and decisions of the rais could strongly influence, for good or for ill, the private lives of ordinary kolkhoz members. This sort of influence was recalled by a man named Zarif, a native of the village of Chopolonchi in the former kolkhoz Madaniyat, during an interview on his orchard farm in May 2004:

> Zarif: After school I worked in the kolkhoz as a *tabelchi* [a low-level accountant responsible for elementary bookkeeping] in 1953–58, and then I went to the army for four years. When I came back, from 1962 until 1964 I worked again as a *tabelchi*. But then they neither gave me a house nor let me study. At that time, in 1960, Qodirhayit was the rais here. He didn't give me the *spravka* [document] to go to study, nor, when I asked, a house. So I left and went to the city to become a construction worker [*stroytel*]. It was 1967 when I decided to leave the kolkhoz with my wife and children. I took the land here

in Madaniyat after going on pension. My house is in the city now. Every day I get to my orchard by bike.
TT: At that time, why didn't you get the house and the *spravka*?
Zarif: He didn't give it to me.
TT: Who?
Zarif: The rais.
TT: But why?
Zarif: I don't know. The rais said so.

Clearly, the kolkhoz chairman's influence was not limited to the sphere of employment but could reach up to influence Zarif's chances of studying and getting a house in his village. In Uzbekistan, the rais, generally considered to be invested with paternalistic responsibility for the needs and livelihoods of his people, had a say in virtually every situation within the kolkhoz's territory that diverged from the routine. Within the limits of Soviet law, he might control his territory as a sort of feudal estate. The rais's political destiny depended on his ability to meet the production targets set by his superiors at the district level. To be kolkhoz chairman was often the first step in a career in the upper levels of the district and provincial administration or in the party.

Within its territory, which in Khorezm often encompassed two or three villages and some 3,000 to 4,000 hectares of land, the kolkhoz was assisted by only one other institution of the Soviet state, the *sel'soviet*, or village administration. In a conversation with an elderly peasant from the former Madaniyat kolkhoz, who in the past had been an *aktivist* – an unpaid assistant to the head of his village neighbourhood committee (*elatkom*)[78] – I asked whether the head of the sel'soviet or the rais of the kolkhoz had more prestige. Without hesitation he answered, 'Of course the kolkhoz rais. He had everything: tractors, cars, fertilisers, *neft* [petrol], everything. We went to him, he did not come to us'.

In the late Soviet period the sel'soviet belonged to the party hierarchy but lacked economic power. In the rural areas it was the most local soviet (council of workers) of the Uzbek Soviet Socialist Republic, the rural equivalent of the mahalla committee in urban areas. These councils nominally were conceived of as cells of the people's participation in the Soviet state. In practice, their most important functions were to register residents,

---

[78] The word *elatkom* is used only in Khorezm; it is the equivalent of *mahallakom* in the rest of Uzbekistan. It is a sub-unit of the village administration, which in Soviet times was called a sel'soviet and since independence has been called the *fuqarolarning yig'ini*. Colloquially, people in Yangibozor use the term *elatkom* to refer to the head of the neighbourhood committee, because in reality the elatkom is not a committee but simply a person with that function. People similarly referred to the head of the Ispolkom, the district executive committee in Soviet times, as 'the Ispolkom'.

disseminate propaganda, support party activities, and control public order. During the cotton harvest campaigns, sel'soviet activists monitored the adhesion and performances of the kolkhoz workers. In much the same fashion, local officials of the successor administrations, the mahallas and village administrations, continue this duty today. The boundaries of the kolkhoz and the area governed by the sel'soviet sometimes coincided, but not always. Some sel'soviets included more than one kolkhoz within their jurisdiction, and vice versa.[79] A sel'soviet included from two to six sub-units, depending on the size of the village. In Uzbekistan outside of Khorezm, these sub-units were called *mahallakom*, but in Khorezm they were *elatkom*, activists were the *aktivistlar,* and the village chair was the *sho'ro*, as opposed to the rais, the head of the kolkhoz.

The problems of the collective farming system arose long before the end of the USSR: kolkhozes were accumulating huge deficits caused by waste and the misuse of collective resources.[80] Expenses for salaries and social goods were not matched by the budgets that kolkhozes received on the basis of their produce. Under the regime of collective property, employees of the kolkhoz saw little incentive to do ambitious work, and farm managers abused their positions to enrich themselves, following the short-term rationale of remaining in power. The terms used widely to characterise the system's shortcomings are 'soft budget constraints' (Kornai 1992) and 'centralised redistribution' (Konrád and Szelényi 1979), coined, respectively, to address social relations and state paternalism in the socialist economy of shortage and to refer to the creation of social inequalities under socialism. Khorezm was not immune to these problems. In the words of many informants, corruption (*poraxo'rlik*) was a major problem of the kolkhoz system, resulting in the creation of a black market, or parallel economy, the basic principle of which was that officials used collective resources for individual gain, partly as compensation for the system's shortcomings but also to favour some people at the expense of others.

A Local Perspective on Late Socialist Modernisation

As many scholars have recognised, the importance of the cotton scandal lies in the fact that its public prosecutions unveiled the existence of a system of power based on a pervasive web of patronage relations encompassing the

---

[79] This complicated some of my data collection, especially concerning the relationship between demography and agriculture, because demographic data (which pertained to the sel'soviet) and data on land use and crop production (which pertained to the kolkhoz) were accounted for separately.

[80] See Lerman, Csaki, and Feder (2004: 44) for a concise summary of these problems.

various levels of the Soviet state hierarchy (see, for instance, Kandiyoti 2002, 2003a). The investigations uncovered the way this system was intrinsically linked with Central Asia's role as a supplier of cotton for the Russian textile industry. Coinciding with the end of the Brezhnev era in Moscow, the investigations started with the discovery that less cotton had been produced than was officially shown in the producers' statistics.[81] What emerged was that for years the central administration had bought cotton that existed only on paper. By manipulating production figures, the local elites diverted to their own pockets a share of the budgetary transfers from the central government. The accusations against and arrests of hundreds of local cadres that followed soon transcended the judiciary dimension. Amplified by articles published in the newspapers *Pravda* and *Izvestia,* the topic gained much public interest and animated the broader political debate in the years just before the Central Asian republics gained their independence.

In the USSR, the public accusations produced a wave of indignation on the part of people in the 'north' over the corruption evidently rampant in the 'south', and public opinion roundly condemned the Uzbek cadres. Those targeted by the investigations perceived the singling out of the Uzbek corruption case as arbitrary, considering the widespread corruption that existed throughout the late USSR. In the Uzbek national arena the accusers themselves soon ended up being accused of having turned an exaggerated public prosecution into a concealed instrument of repression of essentially loyal and legitimate native elites. In this revisionist critique the arrests and trials of hundreds of local cadres in the 1980s were portrayed as arbitrary and illegitimate acts of subjugation that found their closest parallels in Stalin's purges in the 1930s (Kaxramanov 2000). The newly independent Uzbek government adopted the revisionist position, and today, after the public rehabilitation of former first secretary of the Uzbek SSR Sharof Rashidov, it is a consolidated creed in the ideology of the independent state.

Methodological difficulties such as data unreliability, as well as political pressure and preconceptions, complicate the analysis of these discourses today. Irrespective of whether 'justice' or 'repression' was at work, the investigations uncovered how corruption built political consensus through clienteles, created links of solidarity, maintained public order, and succeeded in moving resources from the centre to the cotton-producing peripheries (Buttino 2004). Although these considerations are valid in general terms, today little is known about how this late socialist period affected the local-

---

[81] The investigations led to the imprisonment of hundreds and, with the incrimination of Brezhnev's son-in-law, reached the highest echelons of the Communist Party. The discovery of the fraud was made possible by the use of satellite images of the cotton fields in Uzbekistan.

level arenas in which, after all, the cotton was produced. In addressing these questions the crucial problem is the availability of sources and their use. In the light of the judicial proceedings, the official statistics appear to be of doubtful reliability, just as the published books and statements are likely to reflect the biases of antagonistic discourses.

The ambiguity of sources also applies to national statistics and production targets, which had a twofold, political and scientific, quasi-normative character, because they were meant to predetermine the economic programming of the subordinated territorial units (regions, districts, and kolkhozes). Cropping patterns, work processes, and production outputs ideally had to follow uniform ratios determined by norms formulated by national agricultural research centres. As a consequence, agricultural development was accompanied by tremendous production of data. But what can these data tell us today? The problem with the national statistics is that in the Stalinist tradition of manipulation and distortion, they are highly unreliable and constitute a challenge to correct interpretation.[82] Yet an engagement with the topic of late socialist agricultural evolution cannot ignore the existence of a meticulous statistical data production.

The District Archive

On the outskirts of the district centre of Yangibozor, surrounded by cotton fields, the district branch of the Khorezm State Archive is located in an inconspicuous two-storey building dating from the 1980s. Half the building is occupied by a pharmacy, and half is filled with some 80,000 documents gathered from nearby administrations, enterprises, and organisations (in Uzbek, *tashkilot*). In this building, from spring until autumn 2004, I enjoyed the assistance of the archive staff while collecting data with the aim of gaining a new perspective on events of the recent past in the district and shedding some light on the ambiguous role of the local Khorezmian elites during late socialism, in respect to their relationship towards the socialist state and their local communities. I focused my archival work on documents produced in and by the kolkhozes and sovkhozes, which later became *shirkats*.[83] The problems I have mentioned concerning the reliability and

---

[82] The most sensational case probably is that of the adjusted population figures of the 1939 census, in which the official figures exceeded the real population by 15–16 million, a figure that remained unquestioned until the late 1980s. See Conquest (1991: chapter 14).

[83] The documents I studied were of three kinds: production plans (*moliyaviy ishlab chiqarish plani*), which were compiled at the beginning of each year; annual balances (*yillik hisoboti/otcheti*), produced at the end of the year; and *prikaz* books, or books of orders. The annual balances contained the detailed bookkeeping and statistics for the overall economic figures of the kolkhozes, a balance of their use of land and their overall stock, and information

accuracy of collected data are not specific to the Yangibozor archive but are consistent with the general conditions of Soviet (and post-Soviet) data production and management.[84]

A distinction between the aggregated data produced for official use by the statistical committees and the locally produced, unaggregated 'working data' of the collective farms that could be accessed in the archive helps shed some light on the trustworthiness of the statistics. The aggregated data were the more exposed to manipulation. Because they were used for official purposes, they were often made to fit given target figures in order not to displease the superiors. This was true for unaggregated data as well, but to a lesser degree. Humphrey, in her work on a kolkhoz in Buriatiya, engaged extensively with the problem of the reliability of data produced and used at the local level, apparently using the same types of documents I was able to access in the archive in Yangibozor. She concluded that it was 'probably fair to say that all units, and therefore to a greater degree all farms, operate in an atmosphere of approximation' (Humphrey 1983: 199).

The question is to what extent this approximation is meaningful. It seems to me that rather than denying the data any significance, it is important to understand them within the context in which they were produced. Although the collected figures of the kolkhozes have limits and imprecisions, this does not automatically void them of any worth and usefulness. Interviews with the staff of the archive and sometimes with the authors of the data convinced me that although the figures were approximate, in most cases they were fairly reliable. In the Yangibozor kolkhoz papers, some data adjustments were easy to identify. Sometimes they could be explained by the necessities and variations of the agricultural process or by the disorganisation in the kolkhozes. Sometimes they could be attributed to unfulfilled staff duties. Missing figures or data sets constituted possible evidence of attempts to screen problematic figures.

Comments on these data by Matyoqub Sherjonov, the director of the Yangibozor district archive, were of priceless help to me. Sherjonov was an expert on the local history of Yangibozor, and his soon-to-be published book on Yangibozor revealed a life dedicated to the collection of local stories and

---

about the employees and their families. From them I extracted three kinds of information: data on the administrative links and yearly composition of the staff of the kolkhozes, sovkhozes, and shirkats; data on annual production figures; and data on population and land allocation. I did not study the detailed protocol books of the kolkhozes, which would have been too time consuming. I did consider the *prikaz* books of the two sovkhozes, because they were easier to handle, being shorter and entirely machine written, which was not the case for most kolkhoz protocol books. I have described my archival work more thoroughly in Trevisani (2007).

[84] On the making and unmaking of statistical data in the Soviet Union, see Blum and Mespoulet (2003).

memories of his native district. Since he had come to the archive, his work day had been divided between mornings spent at his desk in the archive and afternoons dedicated to collecting oral histories in the villages of the district and interviewing local personalities and *rayon* elders. Sharing an interest in local history, we occasionally went together to interview people, attend events, and visit places in the district bearing significance for our work. What is sometimes missing in the archived statistics is the 'real life' that Matyoqub Sherjonov told me about, contradicting the writings of other local scholars. His locally informed insights contributed a great deal to filling this gap in my data. Often, when I asked about some unknown acronym, some incongruent piece of data, or some missing folder (*fond*) in which archival documents were stored, his answers led to long digressions into local stories and personalities. We would end up discussing an altogether new subject, inevitably ending with the statement, '[Our] work is infinitely complicated [*ishlar nihoyatda murakkab*]'.[85]

Socialist Development in Yangibozor

Established in 1950, Yangibozor was among the last districts in Khorezm to be introduced as an autonomous *rayon;* it was created by putting together the peripheral areas around the neighbouring larger historical centres of Urganch, Shovot, and Gurlan. In 1958 the administration of the territory was passed to the *rayon*s of Urganch and Gurlan. In 1981 the district was re-introduced, and with the exception of one year (June 1988 to June 1989, when pre-1981 delimitations were re-installed), it has lasted until today.

---

[85] The historical horizon of my archival work was determined by data quality and availability. Manuscripts and documents older than 1924 were stored in the State Archive in Tashkent and were out of reach to me. Data going back to the collectivisation period were stored in the central regional archive in Urganch and needed special permission. In the Yangibozor archive, data referring to the period before and during World War II seemed to be incomplete; documentation on the kolkhozes started to become roughly complete only from the late 1940s onwards. Because of this, and because of the frequent merging and reorganising of the district's kolkhozes in the years between early collectivisation and the late 1950s, I looked systematically at data on the agricultural evolution of Yangibozor from only 1960 forward. The Yangibozor state archive has neither a general catalogue with a description of the documents nor a guidebook to the documents, but instead a rather confusing folder registry (*fondlar ro'yxati*). Every folder (*fond*) for an organisation (*tashkilot*) is listed in a separate sheet. This makes it difficult to get an overview of what is there, although it is easier to follow the documents of any single organisation, if they are listed in the same sheet. For this reason I privileged a systematic view of certain kinds of documents, rather than the documents of different organisations.

Despite these administrative fluctuations, the kolkhozes of Yangibozor remained constant units throughout the late socialist period.[86]

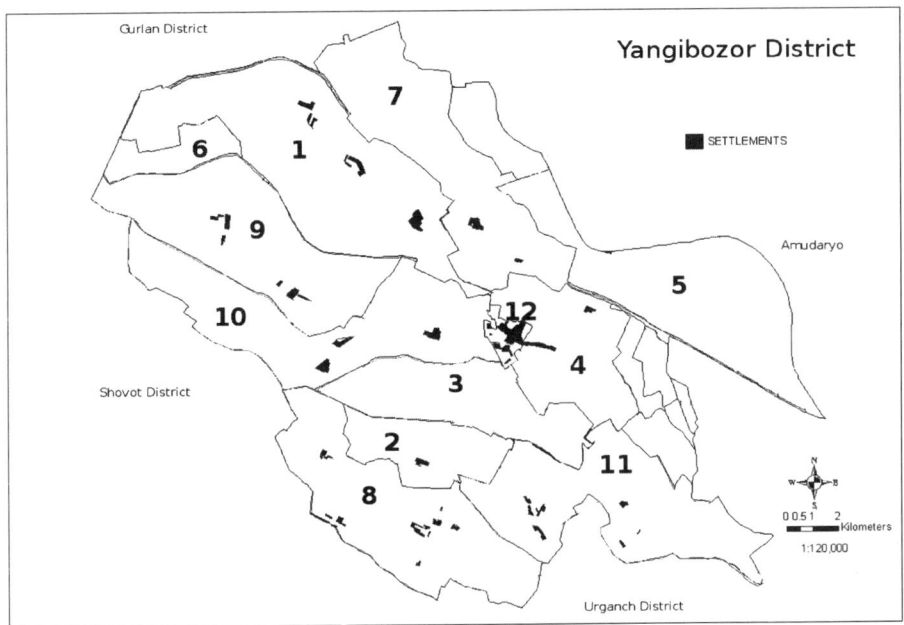

Map 4. The collective farms of Yangibozor district. 1, Bog'olon; 2, Bo'ston; 3, Hamza; 4, Xorazm; 5, Jayhun; 6, Sanjar; 7, Madaniyat; 8, Xalqobod; 9, Bo'zqal'a; 10, Shirinqo'ng'irot; 11, O'zbekiston. Yangibozor city (number 12), the district center (*raytsentr*), is not a collective. Adapted from ZEF/ UNESCO Khorezm project.

---

[86] For the years from 1936 (the oldest mention of a kolkhoz in the archival documents I saw) to 1962, I found traces of the names of 28 kolkhozes. By 1960, after a process of aggregation, all agricultural activity in the district pertained to the eight 'historical' kolkhozes of Yangibozor, with one exception, that of a kolkhoz dissolved in 1962 to be merged with a bigger one. These eight, under changed names and transformed into shirkats, existed until 2002. They were Kuybyshev (later, as a shirkat, called Bog'olon); Leninizm (Bo'zqal'a); XXI Parts'ezd, or 'Twenty-first Congress of the Communist Party of the Soviet Union' (Xalqobod); Oktyabr XIVnchi yilligi, or 'Sixteenth Anniversary of the October Revolution' (Hamza); Moskva (Xorazm); Madaniyat (Madaniyat); Leningrad (Shirinqo'ng'irot); and Pravda (O'zbekiston). After 1960, changes became rare and units more stable. From 1960 to 2002 the only substantial novelty was that two sovkhozes were newly established, one in 1966–1967 (Oktyabr 50inchi yilligi, or 'Fiftieth Anniversary of the October Revolution', later Bo'ston) and the other in 1976 (Rossiya, later Jayhun). One last, very small kolkhoz (Sanjar) was created in the early 1980s to specialise in fodder, although in some respects it remained subsidiary to the largest kolkhoz of the district, Kuybyshev-Bog'olon, from which it split.

On the north-east, Yangibozor district borders the Amudaryo, whose fluctuating banks also delimit the administrative borders of the Karakalpak Autonomous Republic, an autonomous region within Uzbekistan. Several hundred hectares of the former sovkhoz Rossiya (later the shirkat Jayhun), which today is situated on the left bank of the Amudaryo, were once part of the Karakalpak Autonomous Republic, but after a change in the riverbed they have been granted as a perpetual concession to Yangibozor district. Once, a buffer of 10,000 hectares of *to'qay* forest existed between the cropped areas and the 'unruly' riverbed, reaching from Cholish in Urganch district to Gurlan district.[87] In Yangibozor, most of the forest fell to the expansion of the district's agricultural area, which began in the mid-1970s and was accomplished within a few years.

The deforestation went hand in hand with the establishment of the rice-growing sovkhoz Rossiya in 1976, for which many Koreans from neighbouring Urganch and Gurlan districts were employed.[88] A year after the establishment of the sovkhoz on about 4,000 hectares of deforested land, the forest area was reduced to 765 hectares.[89] By 1981 this area had shrunk to only 50 hectares.[90] Besides deforestation, agricultural expansion had other repercussions: all five major lakes of the district were drained and turned into arable land.[91] The expansion of the agricultural area reached its peak in the 1980s.

As in the whole Khorezm region, so in Yangibozor district the intensification and expansion of agriculture during the late Soviet period left heavy traces on the landscape. Even more than collectivisation in the 1930s, the appearance of tractors, pumps, and heavy machinery, which became pervasive from the 1960s, marked a strong discontinuity with the pre-existing landscape by totally redefining waterways, irrigation techniques, and land management. During the late socialist decades an irrigation system of primary, secondary, and tertiary supply and drainage channels was built in

---

[87] *To'qay* is a form of local forest-like vegetation, characterised by its density and impenetrability. Battersby (1969: 18) described to'qay by saying that 'strips of low-lying greenery, trees, shrubs, and vines, like miniature jungles, give colour to the otherwise drab landscape'.

[88] Rossiya sovkhoz was not mentioned by Songmoo (1987: 90), who otherwise reported accurately on the Korean rural communities of Khorezm, that is, on the rice-growing sovkhozes in the neighbouring districts of Gurlan and Xonqa.

[89] Source: Yangibozor District Branch of the Khorezm State Archive (hereafter Yangibozor district archive), Folder 350, Inventory 1, Unit 4 (hereafter given as, for example, F350/I1/U4).

[90] Yangibozor district archive, F350/I1/U12.

[91] The largest lake was Devonqo'l, which extended over more than 1,000 hectares, 50 per cent of which lay in the territory of the district. For the history of the lakes of Yangibozor, see Sherjonov (2003).

order to feed large plots of levelled agricultural land (*kontor*s, averaging 5–10 hectares), a feature that still characterises the Khorezmian agricultural landscape (Veldwisch 2008). In the past, plots had been much smaller and irregularly shaped, and only a fraction of today's agricultural land could be taken, with much effort, from the dry and unarable surroundings, where cattle breeding was the only possible economic activity.

Except for a strip of land along the river, where cultivation is dangerous because of the unpredictable and unsteady river flows, fallow land has disappeared since at least the mid-1970s. Until then half of today's district had been either forest or lakes, which were dried up or cut down and then levelled to be used for rice and cotton production. In those years of agricultural expansion, virtually everyone was employed, directly or indirectly, by the eight kolkhozes and two sovkhozes or in the district administration.[92] During this period of agricultural expansion the government invested in streets, irrigation canals, and machinery (tractors, pumps), in the up-scaling of production and processing, in social services, and in the industrialisation of agriculture.

National Policies and Local Developments

First presented in the resolutions of the March 1965 plenum of the Communist Party Central Committee, the new directives for the intensification of agricultural development would lead, in the two decades to follow, to an unprecedented increase of agricultural production all over the Soviet Union and, in Central Asia, of cotton production especially. The promotion of agricultural development was accompanied by an effort to enhance the 'living and cultural standards' of the rural inhabitants, meaning a pervasive transformation of the character of the peasants' work and of rural life. The kolkhoz worker was to be converted into the rural equivalent of the urban industrial worker (see Kitching 1998). This was to be achieved by turning the organisation of agriculture into something more similar to that of socialist industry and by bringing the living standards of rural people closer to those of city dwellers. In other words, the ultimate goal was the transition of agriculture to modern industrial methods.

In rural Uzbekistan, the state began to make large investments even before 1965 in the use of machinery and chemicals in agriculture and in promoting large-scale land reclamation schemes. Later, with the introduction of a range of new administrative bodies and organisations, the institutional

---

[92] Here I see a difference with the Ferghana Valley as described by Rasanayagam (2002b: 103), because in peripheral Yangibozor, fewer opportunities existed to earn income 'on the side' ('*levyye dengi*'), informally and outside of agriculture.

landscape of agriculture was 'improved' by the creation of an agricultural service sector. Along with it, late socialism expanded the social group of 'white collar kolkhozians' (Fitzpatrick 1994: 140), people raised in the kolkhozes and educated in the cities who then returned to their places of origin to work in the agricultural bureaucracy. In the following years, Soviet literature boasted statistics on traditionally weakly developed rural Central Asia that displayed rising numbers of school pupils, college students, hospital beds, and so forth. Statistics on agricultural production proclaimed a vigorous reform outcome: 'During the decade that followed the March 1965 plenum of the Central Committee agricultural gross output in Uzbekistan increased by 150 per cent, fixed assets by 300 per cent, the power-per-worker ratio by almost 50 per cent. The republic turned out 42,800,000 tons of raw cotton, that is, over 65 per cent of the amount produced by the whole country' (Tulepbayev 1984: 192–193).

The archival data on cotton production in the kolkhozes of Yangibozor confirm the general trends in cotton production at the national level over the same period (see tables in Appendix). Over three decades, the eight 'historical' cotton-growing kolkhozes of the district displayed a small but steady increase in hectares grown in cotton. In 1960 they had, altogether, 8,025 hectares of land grown in cotton. In 1970 this figure rose to 9,034 hectares, and in 1980, to 9,410 hectares. At the end of the 1980s the agricultural area allocated to cotton was still growing. In 1990 the kolkhozes of Yangibozor allocated 10,318 hectares to the cultivation of cotton, 1.28 times the area covered in the 1960s.

Although significant, these figures appear modest in comparison with the overall growth of agricultural land in the district over the same period. In 1960 the eight kolkhozes encompassed 11,745 hectares devoted to their main crops – those mandated by the state, excluding household plots granted for private use – a figure that remained roughly the same through 1970 (11,721 hectares). By 1980 this area amounted to 14,925 hectares. In 1990, with the sovkhozes and kolkhozes fully developed, it reached 19,484 hectares, an increase of 66 per cent over 1970. At the same time, in connection with the new land reclamation schemes leading to the establishment of the two sovkhozes, cotton's share of the district's total agricultural area rose from 68 per cent in 1960 to 79 per cent in 1970, followed by decreases to 63 per cent in 1980 and 53 per cent in 1990.[93] Indicative of the increasing industrialisation of agriculture, the kolkhozes of Yangibozor sharply increased their output of raw cotton per hectare over the years after 1965. During this pe-

---

[93] Percentages are author's calculations based on data from the Khorezm State Archive, Yangibozor District Branch.

riod, according to the archival data, the kolkhozes of Yangibozor stood out for performing well above the national average.[94]

The Collective and the *Rais*

Regardless of their accuracy, statistics from the Yangibozor collectives regarding achievements in production and modernisation conceal the local conditions and difficulties under which those achievements were attained. The official language did not mention the hardships and challenges that attempts to standardise and industrialise agriculture were creating locally. But in the memories of those involved, the achievement of centrally set targets takes on a different flavour. As much as in the early years of Soviet modernisation, prescribed targets were to be met at any cost, regardless of local constraints. And not every kolkhoz leader was endowed with the ability to implement the modernisation policies successfully.

The story of Tojiboy Matsafoyev, from Madaniyat kolkhoz in Yangibozor, well illustrates these circumstances. The father of a rais with whom I worked in 2004, Tojiboy became rais of his kolkhoz at the age of only 35, after being educated as an agronomist at university. After serving as rais from 1969 to 1974, he moved on to the oblast machinery station (*viloyat texnika ta'minot birlashmasi*) in Urganch, where he served as head accountant until 1978. After that he worked as inspector-accountant (*revisor buxgalter*) in the oblast-level inspection commission (*viloyat kontrolne revizionnoye upravlenie*). He died in 2004, and at the mourning ritual traditionally held on the seventh day after a person's death (the *yetti kunlik*), a colleague of his privately recalled the talents of the former rais: 'All that we needed to build in Madaniyat we got from him. There was the plan. But in the plan it wasn't written how to put things into practice, just to put them into practice. The plan said to do such-and-such, but where to get the machinery from, how to get it, this the plan didn't tell [*Plan shunday dedi, lekin technika qayerdan? Qanday? Demadi*]'. Tajiboy's abilities as a provider of building materials and machinery won him the respect and appreciation of his kolkhoz. The story suggests the challenges faced by the local leadership in the collectives and the 'loneliness' of the executors of the plan. The capable,

---

[94] In 1968 cotton output reached an average of 2.50 tonnes per hectare nation-wide but an average of 3.47 tonnes in Yangibozor. In 1980 average output was 3.32 tonnes nation-wide and 4.01 tonnes in Yangibozor. Productivity, measured in tonnes of raw cotton per hectare, rose steadily until the mid-1980s. Calculating from the averages given in the Appendix, tables 5, 7, and 9, in the period 1960–1964 the productivity of cotton per hectare averaged 2.43 tonnes. The figure grew to 3.0 tonnes in 1965–1969, 3.52 in 1970–1974, 3.59 in 1975–1979, and 3.8 in 1980–1984. In 1983, the year with the best results, the district harvested, on average, 4.2 tonnes of cotton per hectare.

skilful *raislar* were those who were able to put the state's orders into practice, despite the difficulties posed by their implementation.

No written source I came across in Khorezm better illustrated the code of conduct implicit in the position of kolkhoz rais – the honours and burdens of *raislik*, the wielding of a rais's tasks and duties – in the late socialist years than the autobiography of Marks Jumaniyozov. Born in 1938, the son of a tractor driver, Jumaniyazov was a native of Oyoqdo'rman village in Yangibozor district and a popular, highly respected former *hokim*, or governor, of Khorezm *viloyat*.[95] After passing through all the steps of the agricultural command hierarchy in his kolkhoz, beginning as head engineer [*bosh injiner*], from 1975 onwards Jumaniyozov ruled over first Xonqa *rayon* and later Urganch *rayon*. After independence he became the regional hokim for Khorezm and later ended his career as a minister in Tashkent. His memoirs (Jumaniyozov 2008) are a mine of information about events and personalities in late socialist and early postsocialist Khorezm.

In narrating his experiences in the kolkhoz hierarchy, Jumaniyozov represents, if not celebrates, a certain style of leadership. The qualities of a good kolkhoz rais are a central theme of the book, and especially in the later chapters Jumaniyozov implies that ruling a *rayon*, or district, was like ruling a kolkhoz 'writ large'. On many pages of the book he characterises the rais as a strong but benevolent ruler, whose strength was linked to his good relationship with his people. In one passage, before he takes power in Xonqa *rayon*, he receives advice from his predecessor on how to govern the district (Jumaniyozov 2008: 159):

> Remember, even if you use a low voice they will do what you have ordered them to do. If you say that you will do so, I have three pieces of advice for you. First, don't say 'they don't work' to those who were close to me, and don't give offence to them. They are a little district; they will not cross the boundaries by themselves. If they don't work let it be so, as long as they don't bring you trouble or hinder your work. Second, don't fight with your subordinates. You have power [*kuch-qudrat*]; they have not. Therefore, they will not forgive. If they forgive, the people [*xalq*] will not. If the people will, God will not. Third, don't seek or keep the company of those who are not respected by the people, as they will make you lose the respect of the people.

This lesson on what made a good district chief emphasises the proximity between the ruler, his subordinates, and the people, as well as their

---

[95] Levin (1996: 167) pointedly portrayed Jumaniyazov as 'a contemporary version of an old-style charismatic leader – the sort of leader who brought to Uzbekistan a form of honest totalitarianism that many Uzbeks seemed to miss'.

harmonious relationship. In another passage, Jumaniyozov describes a good leader as someone who is able to understand the inclinations of his subordinates and who can maintain good working relationships (Jumaniyozov 2008: 164):

> Good leaders don't just lie on the floor. To find them, to raise and grow them, is not an easy task. When you arrive in a new place, someone likes the job he has; another is uncomfortable with the assigned tasks. Before you find out [which is which], much time must pass. Therefore, up until now, wherever I had to go, whatever position I had to take, I have categorically avoided bringing my own group of trustees [*o'z komandam*] or my close relatives and friends [*oshno-og'aynilarim*] to work with me. I don't sack good working people from their jobs. I continue working with those promoted and put into jobs by the rulers who preceded me, even the drivers, the secretaries, and the assistants, unless, of course, they want to quit by themselves.

Again, the virtuous aspects of raislik are portrayed, but it is not difficult to imagine another side of the story, which does not appear in Jumaniyozov's account. The prosecutions leading to the cotton scandal showed how widespread was the negative side of raislik, including the leaders' habit of privileging their own trustees and relatives, regardless of their merit. Since then this habit has gained even greater notoriety, overshadowing the virtuous form of leadership portrayed in Jumaniyozov's narrative.

Returning to Yangibozor, it is not easy to link these 'lessons' to the evidence in the district archive. It has been argued that the 'systematic use of price incentives', such as the relatively high procurement prices of the 1960s and 1970s, rather than 'coercion or administrative fiat' (Khan and Ghai 1979: 26), was the main factor behind the sudden rise of Uzbekistan's cotton production, because it made cotton-growing more attractive to the collectives. Rumer (1991: 70) maintained that 'it was still possible to keep pace with plan targets in the 1960s and, apparently, even in the first half of the 1970s'; it was only later that the 'excessive exploitation of the soil – by ignoring proper crop rotations, by making excessive use of harvesters, and by overusing pesticides – finally began to take its toll'. Yangibozor's district archive contradicts this relaxed image of the 1960s, but it confirms the validity of a strong correlation between the appointment of chairmen and the economic results of the collectives.

As an example, the annual reports of the Madaniyat kolkhoz for the years 1962 and 1964 state that although the targets for cotton were met, those for other crops and for cattle-breeding were not.[96] For the same years the archive reports that the chairmen of Madaniyat kolkhoz were dismissed and replaced, as would happen again twice later in the decade. The situation of Madaniyat was perhaps exceptional. In no other kolkhoz in the district was the chairman replaced so often in the 1960s. The years between 1960 and 1969, however, show a clear correlation between poor economic results and frequent staff replacements, in comparison with the following two decades. Between 1960 and 1969, 11 kolkhoz managers were discharged in 7 of the collectives that would later be joined into Yangibozor *rayon*, whereas there were 9 changes in 9 collectives during the 1970s and 11 changes in 11 collectives in the 1980s.[97] These data suggest that in this less developed area of Khorezm, the 1960s were a particularly harsh and demanding time for the leadership of the kolkhozes.

It seems plausible to say that, in the case of Madaniyat, the social reality that Soviet modernisation policies brought about locally was not always the one desired by the beneficiaries. A closer look at the seed- and vegetable-growing sovkhoz in the district, which was called 'Oktyabr 50inchi yilligi', or Fiftieth Anniversary of the October Revolution, and which became the shirkat-MTP Bo'ston, shows that it might not even have been the social reality intended by the planners. Allambergan, the elatkom in Bo'ston at the time of my fieldwork, described the establishment of his sovkhoz in 1967 in a thoroughly positive way:

> The sovkhoz was made by putting together the three worst brigades [meaning land, not people] of the neighbouring kolkhozes. They had no streets, they had salty soils, they were the worst. Out of 1,000 hectares of land, maybe only 150 to 200 hectares were irrigated. At that time, Bektosh Rahimov, from Samarkand, was *obkom* [head of the oblast] of Khorezm. They took the three worst brigades, put them together, and transformed them into the most beautiful place: they brought the street, canals, houses, new [irrigated] land.

The book of orders (*prikaz daftari*) of the sovkhoz Oktyabr 50inchi yilligi for 1967–1993 offers a more variegated picture of the issues and problems emerging in daily reality during the late socialist period of agro-industrial development. Electricity arrived there in 1968. Natural gas for heating and cooking was installed during the winter of 1978–1979. Industrialisation, however, never came. The *konservzavod,* a vegetable processing

---

[96] The annual report of 1964 said, 'The low profitability of cattle-breeding resulted in a deficit of 40,028 *so'm* [roubles]. The plan of the kolkhoz to obtain 940,870 so'm was accomplished to 89 per cent, or 844,156 so'm. Accordingly, instead of the 465,525 so'm foreseen for the payments to the kolkhoz workers, only 340,341 so'm were paid out'.

[97] Author's calculations based on data in the Yangibozor district archive.

factory (the only factory in the district except for the cotton ginnery), finally opened on 9 August 1991, after many postponements, but it never worked properly and quickly closed down under the difficulties of early independence.

From the entries in the book of orders it appears that the rais of the sovkhoz was permanently busy keeping his subordinates on track and overseeing production through ad hoc committees, for which he was occasionally compelled to call his trustees back from vacation or business trips.[98] The ideal of a modern, industrial organisation of labour appears only on paper, where the prikaz entries give a semblance of the language and relations of bureaucracy. Typical entries regarding the everyday administration of the sovkhoz pertain to issues such as the release and appointment of staff, granting of holiday permissions, provision of welfare services, and, occasionally, the awarding of cash prizes such as those given to the 40 winners of the socialist competition among pupils of the local school in 1982.[99] All orders were stamped and signed by the sovkhoz director and the legal advisor. Yet the concentration of important positions in the hands of the few educated, capable, privileged persons (who in interviews were mentioned as the *kerakli odamlar,* the 'necessary persons') and the 'dynastic' transmission of top positions from father to son show, on the contrary, that the collective maintained the features of a personal and familial network, antithetical to its claimed modernity.

The revenues obtained by the collective through plan fulfilment did not reward the kolkhoz members but increased the reputation of the chairman. On the occasion of the 160 per cent fulfilment of the production target at the sovkhoz in 1980, the surplus income was put into a 'fund of financial encouragement'[100], which enhanced the rais's possibilities for redistributing resources. He decided whether brigade leaders would be granted cars and other prizes.[101] In the prikaz books, the rais both receives and confers prizes.[102] He grants land for private use to particular individuals[103], and his right to do so, as first representative of his community, is neither questioned nor challenged. The books show how, when the community celebrated Soviet festivities such as May Day, it did so in a traditional fashion, with the

---

[98] Oktyabr 50inchi yilligi sovkhoz, order no. 33, 11 March 1978, and order no. 28, 5 April 1973.
[99] Oktyabr 50inchi yilligi sovkhoz, F7/I1/U69, order no. 77, 30 September 1982.
[100] Oktyabr 50inchi yilligi sovkhoz, F7/I1/U62, order no. 37, 31 December 1980.
[101] Oktyabr 50inchi yilligi sovkhoz, F7/I1/U37, order no. 31, 20 December 1977.
[102] Such cases were documented in, for example, Oktyabr 50inchi yilligi sovkhoz, 7/1/69, order no. 42, 30 June 1982.
[103] Oktyabr 50inchi yilligi sovkhoz, F7/I1/U32, order no. 9, 3 May 1973.

support of the rais, who ordered ingredients for *palov* and specified their quantity.[104] In the archival papers the image of the rais is that of a scrupulous observer of the rules and a lonely fighter against waste, theft, 'backwardness', and his people's resistance to modernity. In one case, the prikaz tells that the rais has reported to the authorities a husband's mistreatment of his wife.[105] Not infrequently, local staff are fined or fired for incompetence, fraud, or theft.[106] Punishments sometimes have the flavour of personal vengeance, but often they disclose the rais's incapacity to handle local manifestations of structural problems, such as systematic theft, damage, and free-riding on state property.

Because I could not study the protocol books of all the collectives as closely as I did those for Oktyabr 50inchi yilligi, it is difficult to determine how accurate this representation of the issues and problems of a collective farm during late socialism in Yangibozor really is, and how far the image of the rais that emerges can be considered representative. The prikaz books for the other sovkhoz in the district, Rossiya (later the shirkat Jayhun), give much less detail about the administration's everyday concerns than do those for Bo'ston. I did learn from its prikaz books, however, that the rais at Rossiya during the late 1970s augmented the number of the brigades there. Every brigade had to deliver a certain number of rams, because, I was told, the rais was interested in traditional Khorezmian ram fights. When it came to granting prizes, the rais at Rossiya seemed to be more generous with himself and his closest staff than the rais at Bo'ston was.[107] In comparison with Bo'ston, at Jayhun orders for recruiting and dismissing staff were issued more frequently. There, too, many orders concerned the assignment of prizes and of special treatment. Such orders increased in importance throughout the 1980s. The impression given is that obedience to administrative rules varied from collective to collective and that generally the stricter discipline of the early years relaxed over time.

A superficial glance at the protocol books of the kolkhozes confirms this, for it appears that people's participation in the local assemblies steadily declined. Indirectly, this can be judged as a further signal of the unruly growth of bureaucracy. This trend emerges clearly in the following conversation I had with a staff member of the district archive:

---

[104] Oktyabr 50inchi yilligi sovkhoz, F7/I1/U18, order no. 6, paragraph 1, 2 May 1972.
[105] Oktyabr 50inchi yilligi sovkhoz, F7/I1/U34, order no. 79, 12 August 1978.
[106] Such cases were documented in, for example, Oktyabr 50inchi yilligi sovkhoz, F7/I1/U32, order no. 8, 17 January 1973; F7/I1/U32, order no. 9, 3 May 1973; F7/I1/U38, order no. 57, 27 March, 1981.
[107] Rossiya sovkhoz, order no. 5, §5, 7 January 1989.

TT: I saw in the protocol books that at one time there were many persons attending the meetings of the kolkhoz. Is this true?

MK: Yes. In the 1950s there were so many people attending the assemblies of the kolkhoz that they were held outside. There were no buildings large enough. Everybody sat on the ground, except for the rais, who sat on a stool.

TT: When did people stop attending the meetings?

MK: During the 1960s there were already many meetings [*majlis*], but then starting from the 1970s, the number of meetings increased steadily. There were *qvartal* [quarterly] meetings, *sessia* [session] meetings, so in the protocols of the kolkhozes it turned into a *formalizm* with little importance. Before, there were meetings three times a year, but then you had meetings three times a month, and so you had fewer and fewer issues to debate, while the administrative offices [*tashkilot*] dealing with those issues became more and more numerous.

The Cotton Scandal as Seen from Yangibozor

Beginning in the mid-1980s, Yangibozor district entered an economic and political crisis. In 1985 and 1986, seven raislar from 11 collective enterprises were discharged and replaced. In at least one case the dismissal of a sovkhoz director was related to accusations leading to a trial.[108] During the second half of the decade, cotton production declined sharply. In the years 1985–1989 the output of cotton per hectare (3.14 tonnes) fell back almost to the level of the mid-1960s. The area allocated to the cultivation of cotton, however, remained large, suggesting a political determination to keep cotton production high despite lower profitability. During these years, all cotton-growing kolkhozes significantly increased their areas planted in cotton, although they obtained significantly less output per hectare than they had previously (see Appendix, table 9). In 1988 and 1989, the rice-growing sovkhoz Rossiya (Jayhun) began to allocate areas to the cultivation of cotton, a clear sign of the district leaders' apprehensiveness after the poor

---

[108] Rossiya sovkhoz, 1986, order no. 3/97, paragraph 4, 13 June 1986: 'Qo'chqorov Rahimboy, sovkhoz director – removed from position since June 13, 1986. On the basis of order no 63/1 of Khorezm oblast agro-industrial committee (Agroprom)'; paragraph 7: 'Karimov Abdullo, chief accountant, relieved of his position due to neglect of his position and assisting some people in theft. On the basis of order no 25 of the Khorezm oblast Agroprom committee'. Rossiya sovkhoz, 1987, order no. 10, paragraph 1: 'In order to attend at the Xazorasp *rayon* popular court the trial of the former sovkhoz director Qo'chqorov and the former chief agronomist Bobojonov, the legal advisor of the sovkhoz, Qurbonov, is sent in mission to the court of Xazorasp'.

performances of the years before and of their will to ensure meeting their cotton production quota at the district level.

Tellingly, crucial data about the years of the cotton scandal have been lost. Documentation of production at the largest cotton-growing kolkhoz in the district is missing for the years 1981–1989, and the folders containing the documents of the district administration (Ispolkom) for 1980–1986 were apparently lost on their way to the archive. Yet the cotton scandal clearly involved the systematic manipulation of production figures. The literature describing this manipulation says that it inflated the amount of cotton delivered and concealed the expansion of the area under cultivation (Rumer 1991: 70). In my conversations with people who recalled the years of the scandal, it emerged that several times, large amounts of cotton were burned at the cotton collection point in neighbouring Gurlan, which for a period was used by the northern kolkhozes of Yangibozor. When I pressed for clarification, I learned only that 'these were cases for the criminal court [*jinoyat sudi*]' and that in at least one case the cotton 'had been burned in order to hide overproduction, which would have been illegal'.

During interviews, my impression was that such issues were still topical and therefore avoided. My attempts to collect more nuanced local information about the economic, political, and moral facets of the years of the cotton scandal met with disappointment. Matyoqub Sherjonov, the director of the district archive, who had served in the local administration for a long time, recalled the years of the scandal in the district as follows:

> MS: In 1982, starting on January first, Yangibozor was created as a district. Shokir Matniyozov worked there as both Raikom and Raispolkom [head of the *rayon*-level party committee and of the *rayon*-level executive committee, respectively]. In that year the cotton harvest was very good – such *to'y*s were celebrated! I myself wrote many of those statistics.
>
> TT: Yes, I have heard about the good harvests. But then, what was the problem? Why were there investigations?
>
> MS: You see, my friend, that's life. Objectively, all over the world the main struggle is to improve living conditions. ... So in 1982 the climate conditions were very good, not only in our Yangibozor, but all over Khorezm, all over Uzbekistan. The cotton production targets were fulfilled very well, in all oblasts, and among those, in our Yangibozor district. For this reason, the economic potentials of the whole of Uzbekistan, of Khorezm oblast, and of Yangibozor developed. Construction works were made, people built new houses, they made their *to'y*s, many people bought cars, the ones who had a bicycle passed to a motorbike, those who rode a motorbike got a car, and

so on, meaning that this was a gift of God or, if we want to see it from the Marxist point of view, a gift of nature. ...

So, the country developed, the people of the oblast and of the *rayon* strengthened their economies, and in society somehow positive changes began to appear. A country, depending on its political system, can judge such a development differently. In my opinion Moscow was afraid that if Uzbekistan continued to develop in such a way, it would not remain obedient to Moscow anymore. Seeing that a big potential was developing here, they started to fear, perhaps, that an independent state could be on the way. And so, with the proposition of keeping the loyal and expelling the bad elements, the leaders sent Gdlyan and Ivanov with their agenda to check, and they called it the cotton affair, or Uzbek affair. But in reality, they sent them with the clear purpose of bringing trouble.

When they arrived, the first thing was to target our leaders [*rahbarlarimiz*]. 'You have falsified the papers [*perepiska qilib turibsizlar*]; such an increase in the cotton yields is not possible', they said. At that time all of our leadership was subordinated to Moscow, but then, if it was so, from where else could we get the money? 'No, you have cheated on the cotton, you have falsified the registry books, you have stolen from the state'. They started a repression [*qatag'on qildilar*] against all our leaders, at the oblast level, at the republic level, in the kolkhozes. So the people's spirits again fell; everything was going so well before. Our leaders left, they were all jailed.

TT: Can you give some examples of Yangibozor's leaders during those years?

MS: In those years there was a rais called Sobir ('Sovur') Yoqubov, rais of O'zbekiston, he was such an excellent person that the kolkhoz called O'zbekiston lost its name and everyone said 'Sovur Yoqubov kolkhoz' instead. To such an extent the people respected him. He became rais in the 1960s and worked until the 1980s, while today his son is still rais there. He was one among the young raislar who worked with the good motivation of serving his own country. Another one of those was the rais of Leningrad kolkhoz, Erkin Yusupov, he was rais for just three years, but he earned prestige as if he had been for thirty years. You see today also, at the head office of the kolkhoz – recently it was closed down – he had planted 50 *gujum* trees [a local sort of elm tree]. So until recently, when there were meetings [*majlislar*] in Yangibozor, it would have been near the office building of Leningrad. Everywhere you find the *gujum* trees there, giving cool, shaded places. Another of those raislar was Bobojonov Yoqubboy. ... Another was ... the rais of our kolkhoz in Hamza, Safar Quvoq – he was rais for 30 years – and after him Sobir Otajanov, who did raislik for 13 years, but the prestige that the first found in 30 years, the second found in 13.

TT: These were the managers in office when the investigations by Gdlyan and Ivanov started. What did the people say about these raislar, and what about the accusations against them?

MS: These raislar, they were the people [*shuni xalq*]! And therefore the people of Yangibozor were angry at Moscow. For instance, Sobir Otajonov, Safar Niyazov, Sobir Yoqubov, Qurbon Bobojonov, Erkin Yusupov: the people from Yangibozor, from Bog'olon, Shirinqo'ng'irot, Hayvat, Uyg'ur knew them, they were born and raised among them. ... These were the people they served, they were known by them, but Moscow did not know them, the group of Gdlyan and Ivanov did not know them. Among the Uzbeks there is a saying, 'If your claim is against the khan, to whom will you appeal [*Da'vogaring xon bo'lsa, arzingni kimga aytasan*]?' There is nobody bigger than him to whom can I make my cry. He will hang me! Therefore these people became the victims of Gdlyan and Ivanov. ... They crushed down [*bosib qo'ydi*] the new leaders, who were educated, raised, and ready to rule. What does this mean? It means first of all from the economic side: if the canals have to be cleaned, they will not be. In the same way, instead of the five excavators they had to deliver to Yangibozor ... , they gave only three. Can three excavators do the work of five? They can't, even if they work day and night. Moscow knew this well.

TT: As you said, after the old leaders were removed, a new generation of leaders came. Can you tell a bit more about them? What was the difference with the former generation?

MS: I told you already, may your opponent in court not be a khan [*da'vogar xon bo'lmag'ay*]! The last raislar remained totally within the limits of their competences: what they were instructed to do they did, what they were told not to do they didn't. And so did the people, because they realised that whether they worked or not, in the end it would be the same. As the Russians say, 'You work, you work, you don't get rich, but you become crookbacked [*Rabotaesh', rabotaesh', ne budesh' bogatiy', a budesh' gorbatiy*]'. The Uzbeks saw this in their own lives. With this inclination the new generation grew limited in political and economic respect, as well as in their rights and in their professional skills. 'If an order from above comes I will work, if not I will rest. If they fire me from my position I'll go, I'll become a teacher. If they don't want me to be teacher, I'll stay home and do nothing, what else can I do?' Here you have your sociological analysis. If you look at the archive material from this perspective, you will see how different were the years 1981–82–83–84–85, with the exception of 1986, from the data of the years 1987–90. You can already see how the mentality had changed. In the three years 87–88–89 the condition became very bad, in agriculture, in industry, and in the public-cultural construction works. ...

TT: I have heard that public prosecutors came mainly to check the data of the kolkhoz papers. Apparently, the production figures for 1981, 1982, 1983 etcetera were higher than in reality. Did this happen in Yangibozor?

MS: No, this was not the case. I will demonstrate this to you through the story of these people and their way of life. Today there is a person called Ro'zim Masharipov in Hamza, a person with a very good heart, a very advanced mentality, whom everybody respected. He would tell directly if something was wrong; if it was a lie he would even go against his own father. This person was jailed for one and a half years because of only 16 cents. ... Such kinds of people were fooled by Gdlyan and Ivanov's group; they removed them from their jobs or put them in jail because in their mind those people didn't think about the affair of Gdlyan and Ivanov, they were thinking of making an affair of all society, they were against the whole society, they were against Uzbekistan. This was not the fault of society; it was the mistake of the cheaters of society. At that time the cheaters of society were the leading state organs. If it had been the lower ranks it would have been possible to remove them. But they were at the top, they were Moscow, the khan! Whose power could reach all the way to Moscow?

Therefore, when the time came to reconstruct Uzbekistan, after almost five years of perplexity, he [President Karimov] was very surprised, because in order to re-build a healthy organism, one had first to re-build healthy thinking. To a whole generation it was necessary first to develop a firm and healthy way of thinking, such as 'We were not finished as persons, we have to be conscious of us as valuable persons'. Gdlyan and Ivanov got what they deserved, ... they received God's punishment. But now the men of that period have become old. I myself became old, I was young at that time; Ro'zim aka was a strong agronomist at that time, he is old now. We are the last witnesses of that time.

The Legacy of Late Socialism

Sherjonov's account of the years of the cotton scandal provides some insight into relations between local communities and their elites in Yangibozor and into the remoteness of their language from the official language of the 'centre'. In his account, the kolkhozes of Yangibozor received their demotic names according to the raislar who ruled them. He portrays the people and their rulers as being 'as one'; their proximity is emphasised as much as the injustice of what he terms 'repression from Moscow'. These statements, however, should not be interpreted as reflecting a lack of loyalty to or an alienation from communism, but as additions to the understanding of local perceptions. This local account is not alternative to the official post-independence narrative but follows the furrow of the governmental position on the rehabilitation of the purged cadres. Although it does not add a real specificity to the district (Sherjonov's description would probably fit many other districts of Central Asia that were hit by the cotton scandal), his recalling of the effects and perceptions of the scandal gives voice to the view of the many mid- and local-level officials that were targeted by the prosecutions of those years.

The local approach to the events helped to shed some light on the late socialist kolkhoz, beyond the methodological impasses rooted in problems of data reliability and beyond the rhetorics of the 'accusers' and the 'revisionists'. A first result is that we can see how the generation of rural elites who were charged with accomplishing 'the plan' and promoting modernisation was left alone to fill the gap between local conditions and Soviet norms and planning. Although until the cotton scandal they were not in an uncomfortable position, the raislar's work of mediation between their communities and Soviet policies, and between general and particular interests, was far from easy. However, against the idea that these local agricultural elites were trapped between local constituents and socialist rulers in a dilemma of loyalty (Gleason 1991), the material from Yangibozor seems to confirm that the key question regarding the late socialist development of rural society was about diverging understandings of what it meant to be a good communist leader. It suggests that this divergence resulted from discrepancies between the 'high' notions implicit in state policies and the notions enacted by local state actors through their particular *habitus* of power, in which the categories 'Soviet', 'modern', 'traditional', and 'local' intermingled. Rural communities, with their views and values, influenced the ruling style of their local leaders, and this resulted in conflicting local and official views of what constituted good, legitimate leadership conduct.

In the nuances of the narratives emerging from Yangibozor we see the paradox of local-level and state-level discourses that used the same words but told different stories about what was to be proper socialist development. On one side there was the 'insider' discourse of the Uzbek cadres, sidestepping the contradictions of serving the state and their communities at the same time. On the other side was an 'external' discourse (for example, that of the prosecutors from the centre) that disclosed and condemned the collusive contradictions of the first discourse. In Yangibozor, this distinction was illustrated to me with a statement attributed to Bektosh Rahimov, the chairman (1965–1970) of the association of war veterans of the Uzbek SSR. 'In life there are two types of work', he said. 'The first is that which is mandated by law but which is useless; the second is that which is not mandated by law but which is advantageous for society. If you think of carrying out the work without the law [meaning the work that is good for society], we will find a law for you from the state'.[109]

This sentence embodies the 'philosophy' of the local cadres and their understanding of legitimate rule. With greater discernment of local perceptions of events, one sees a distinction between 'honest' and 'dishonest' forms of corruption among those accused in the cotton scandal. What emerges from this distinction is a locally legitimated model of leadership that contrasts sharply with the accusations of 'localism' expressed by the centre at the time (Critchlow 1991: 141). Ultimately, the kind of relationship between xalq and rais that was established in the kolkhoz, consolidated during late socialism, and manifested during the cotton scandal rested on a particular understanding of state communism as a system of ruling, but also on locally shared ideas of authority and leadership and on a particular centre-periphery constellation. With the transition to postsocialism, elites attempted to hold onto this sort of 'indigenous socialism' and, as I describe later, to adapt it to a new political economy of cotton.

---

[109] 'Hayotda ikki xil ishlari bor. Birinchi qonunda bor lekin foydasi yo'q. Ikkinchi qonunda yo'q lekin jamiyatga foydali. Siz qonunda yo'q ish bajaring dedi biz davlatdan qonun topamiz'. Author's translation.

Chapter 4
Postsocialist Agriculture

Uzbekistan's postsocialist agrarian reform policies have ambiguous traits, which neither the principle of 'transition to the market' nor the formula 'continuity with the past' adequately captures. On the one hand, the policies were aimed at strengthening and strategically reasserting the country's newly achieved independence. On the other, all reform efforts initiated after independence were directed towards more efficient use of resources, increasing the peasants' motivations for working, and cutting off the parallel economy, which undermined the efficiency of the kolkhozes. Rather than a straightforward adoption of liberalism, Uzbekistan's post-Soviet reform story can best be understood as an attempt to overcome the problems and shortcomings of the Soviet kolkhoz.

During the perestroika years, debates over the need to reform Soviet agriculture had already led to the diagnosis that a lack of clearly defined responsibilities, or the land's 'masterlessness', rather than the absence of private land ownership, had caused the crisis of Soviet agriculture, and conversely, that its solution was to be found in the saying, 'The land needs a master' (Shanin 1989: 15). The Uzbek government, perhaps even more than those of other post-Soviet countries, concurred in this diagnosis, and it reformed agriculture by increasing individual responsibilities throughout the production process, without lifting regulations and state ownership of land. Instead of deregulating agricultural production, Uzbekistan has in fact re-regulated it. In the process, pervasive reforms have arisen from each new set of constraints and needs. The reforms have been aimed at modernisation, but in a way that has not disrupted the pre-existing social, economic, and political arrangements of agricultural production.

In this chapter I describe the effects of Uzbekistan's agricultural reforms on one rural district, Yangizobor. I begin by tracing the country's land reforms and changing regulatory framework since independence. I then discuss the ramifications of privatisation and subsequent decollectivisation, and I summarise the effects of the re-regulation of agriculture on the command structure of production. Despite a new institutional framework, in Yangibozor the old command hierarchy has essentially been maintained, and the district authorities use coercion and control to sustain this hierarchy. The outcome is a mixture of continuity and rupture with the organisational structure and power relations of the past. *Fermer*s are the emerging actors in the reformed agricultural sector, but their role is ambiguous and their chances for future success are mixed. For them, good connections with district authorities and access to favourable land leases are at least as important as the 'hardware' of agriculture for profitability.

Postsocialist Agropolicy and Land Reform

Sharing the legacy of a uniform regulatory background, the countries of the former Soviet Union have set off along various paths, adopting different land reforms and degrees of liberalisation and implementing diverse market-oriented rules (Wegren 1998). Amongst these countries, the Uzbek case entails the paradox of a reform that avoided introducing any substantial features of a market economy and liberalisation but nevertheless caused important changes in the agrarian structure and the relations of agricultural production. In order to understand relations between rural communities, agricultural producers, and the state, it is necessary first to understand the complex architecture of decollectivised agriculture. Its technical aspects are all significant for the study of Yangibozor's rural communities.

Lerman (1998: 157) summarised the salient attributes of the reforming Uzbek agrarian system as the 'retention of state monopoly on land ownership and continued pervasive intervention of local and central authorities in agriculture'. While some substantial transformations did get under way, including a new land code and the decollectivisation of the former large state farms and cooperatives (initially towards the formation of much smaller, individual, more commercially oriented farms), 'the combination of state monopoly on land and continued central controls' observed by Lerman (1998: 157) have remained in place. Pervasive state interventions and an economic system based on quotas and production targets have been maintained, so that in this respect, unlike in most of the former Soviet countries, the state's retreat from agriculture has not really started in Uzbekistan.

Under the influence of the Soviet legacy and of the economic shock following the end of the Soviet Union, the priority of post-independence rural policy has been the centralised, commanded production of cotton and wheat, for which local producers have paid a high price in social and economic terms. This priority has been explained by the need to safeguard the country's newly achieved independence economically and by the political determination to maintain self-sufficiency in food, although wheat imports could be economically more advantageous (Kandiyoti 2003b).[110] Cotton is among the most important sources of revenue for the government, which until recently maintained the position of monopolistic buyer and therefore gained a large advantage from exporting this profitable crop, to the detriment of local producers (Kandiyoti 2003b: 146–147, 2009: 24).[111]

International organisations and external observers have pointed out the economic shortcomings (International Monetary Fund 2000) and social costs (Guadagni et al. 2005) of this policy, rooted as it is in the legacy of the past. Yet they also acknowledge that initially Uzbekistan's 'reform conservatism' helped to absorb the economic shock better than the more reform-oriented neighbouring republics did (Kandiyoti 2003b: 143). With a large share of national GDP produced by the rural sector, and with the majority of the population earning its living directly or indirectly in that sector, the restructuring of the agricultural system is of pivotal importance for Uzbekistan (Spoor 2006). The declared goals of the Uzbek reforms have been to reduce the wasteful use of resources, to counter ecological degradation of the agricultural environment, and to reverse the diminishing productivity of the country's large agricultural enterprises, the former sovkhozes and kolkhozes. This was to be achieved by transferring more individual responsibilities and liabilities to agricultural producers and by creating new incentives based on the leasing of former collective land to individuals and families.

Uzbekistan's agriculture features a dual structure (Ilkhamov 2000). The largest share of land is allocated to farm enterprises and is used to grow centrally commanded crops (cotton and wheat), while a small portion is allocated directly to rural households in the form of subsidiary plots, little different from Soviet times. Plate 7 illustrates this dual structure, showing large, state-determined cotton fields juxtaposed with the more attentively cultivated household 'smallplots' of a rural settlement. On these plots, peasants grow rice, wheat, or vegetables, mainly for their subsistence needs. While household plot production is directly available to producers for their own consumption or for sale, the two strategic crops, cotton and wheat, remain governed by the state procurement and planning system. These crops are still largely sold at fixed prices to state-controlled retail stations, grain collection points, or cotton processing facilities, and a substantial part of the generated value percolates away from the grower.

---

[110] On this point, Zanca (2010: 65) writes: 'Relative to greater export commodities, such as gold and oil, cotton requires little capital investment, to say nothing of not having to share profits with foreign concerns'.
[111] The cotton processing and export sector has been gradually privatised in recent years, but the government continues to control cotton revenue, although more indirectly.

Plate 7. A rural landscape in Khorezm, October 2004. On the right are large fields planted in cotton; on the left is a rural settlement (*posolka*) with annexed smallplots.

Uzbekistan's independence brought an end to the Soviet subsidisation of agriculture, and the sector took on a new role as a net donor in the national economy.[112] Although Uzbek law promises a certain degree of freedom, in fact producers are still compelled to sell all their cotton to state ginneries at prices fixed by the government below the world market price.[113] Consequently, their profit margins are considerably reduced while the government can make a profit in international currency. Rather than direct taxation of land and water, the government's maintenance of a crop procurement system, coupled with its control of exports and regulation of local

---

[112] Scholars still debate the degree of cotton subsidisation, and because of the non-transparency of the cotton economy figures, experts have come to contradictory conclusions (cf. Müller 2006; Guadagni et al. 2005).

[113] In recent years, cotton ginneries have been privatised. In Yangibozor, for instance, the cotton ginnery had been turned into a company owned by private shareholders. Their economic freedom, however, was constrained and, much as in the past, subordinated to the directives of the command-administrative hierarchy.

prices for cotton, has created the conditions for high indirect taxation of agriculture (see Kandiyoti 2003a, 2003b; Trushin 1998).[114]

During the first decade of independence, the Uzbek leadership drew resources from agriculture to finance its import-substitution industrialisation (Ilkhamov 2000; Trushin 1998; Guadagni et al. 2005). This practice accentuated the problems of the dual agricultural system and aggravated the postsocialist crisis in the agricultural sector. Post-independence agriculture was characterised by comparatively poor performance, a result of weak incentives for individuals to enhance the productivity of land sown in 'state crops'. Money-losing enterprises became unable to pay wages, impoverishing the workers on former kolkhozes. Supplies of agricultural inputs and the retailing of produce had to be reorganised along new domestic channels, which created bottlenecks, especially in the availability of fertilisers and fuel. The lack of investment in a deteriorating capital stock, especially tractors and other machinery, much of which had been acquired in the 1980s and not renewed since, increasingly 'de-mechanised' agriculture and essential irrigation infrastructure, slowing the development of agriculture.

The government's agricultural policy must be understood as an attempt to cope with these difficulties. On one hand, in the early years of independence the government strengthened household subsistence by doubling the area of the household subsidiary plot, from 0.13 to 0.25 hectare (Spoor 1993). To do this, it redirected a portion of the area cropped under the state-ordered system to households. On the other hand, it gradually transformed state crop agriculture from the collectivist organisation of the past into a new organisational form, the *shirkat,* based on individual contracts within newly formed joint stock companies. The reforms, however, did not include the privatising of land. Land remained under the unalienable ownership of the state, although the reforms did entail a redefinition of the conditions of use for producers. This process was accompanied by a restructuring of the large, state-controlled farms, the creation of new types of autonomous enterprises based on leasing agreements, the reorganisation of the agricultural supply and retail structure, the introduction of water user associations (*suvdan foydalanish uyushmalari*)[115], the introduction of a new

---

[114] Land and water taxation still had minor importance relative to the indirect taxation represented by compulsory production targets. Water taxation, for instance, was calculated on the basis of the area of the leased land, and not on real consumption in cubic meters. Both land and water taxes were so low that they mattered little in the balances of the farms I dealt with. It appeared, however, that land and water taxation would grow in importance in the future and replace indirect taxation.

[115] At the time of my fieldwork there were water user associations (WUAs) in every district, but they were administrative entities and did not follow hydrological boundaries. Instead, a

land cadastre, and the beginning of a gradual lifting of market constraints on pricing and cropping decisions.

Shirkat, Fermer, Dehqon

The most important measures of agrarian reform were introduced with the law on farmer (*fermer*) and peasant (*dehqon*) enterprises promulgated in April 1998.[116] With this law, preceding laws and decrees were reorganised, and the national agricultural system was re-established around three types of farms: joint stock companies (shirkats), fermer enterprises (*fermer xo'jaligi*), and dehqon enterprises (in Western literature sometimes called *dehkan* farms or peasant smallholdings). The new land code reaffirmed state ownership of all land (article 16, part 4), defined the types of admissible land use (article 8), and stipulated the jurisdictions of state bodies (the cabinet of ministers, in article 4; the regional government [*viloyat hokimiyati*], in article 5; and the districts [*tuman*s, *rayon*s], in article 6) in regard to land issues. Among other things, it gave the district governor (*hokim*) the exclusive ability to grant land to citizens for farming, defined the viloyat's and the tuman's jurisdictions over the monitoring and control of land use by juridical and real persons (article 14), and determined the modalities for realising state control over the use and protection of land (article 85).

The law granted users the right to life-long and inheritable possession of dehqon land plots and of parcels of land for housing (article 19), but it limited the possession and use of other agricultural land parcels to a form of long-term leasing (article 24) based on a leasing contract (*ijaraviy shartnoma,* or, in short, *ijara* contract) linked to a specific business plan. The ijara contract became the legal basis for fermer enterprises. Complimentary to it was the *pudrat,* a legal agreement defining the working relationship between a head of a rural family and a shirkat (article 51). Most important among the restrictions placed on land users and land leasers (part 5) was the strict

---

WUA could comprise one or more former kolkhozes. On WUAs in Khorezm, see Wegerich (2010) and Veldwisch (2008).

[116] The essential Uzbek legislation on land consists of the following: the Land Code of the Republic of Uzbekistan, no. 589–I, 30 April 1998, last amended 30 August 2003; the Law of the Republic of Uzbekistan on Agencies of Self-Government of Citizens, no. 758–I, 2 September 1993, last amended 14 April 1999; the Law of the Republic of Uzbekistan on Agricultural Cooperatives (shirkat), no. 600–I, 30 April 1998, last amended 12 December 2003; the Law of the Republic of Uzbekistan on dehqon enterprises, no. 604–I, 30 April 1998, last amended 23 May 2005; the Law of the Republic of Uzbekistan on Farmer Enterprises, no. 602–I, 30 April 1998; and the Decree of the President of the Republic of Uzbekistan on the Plan for Development of Farmer Enterprises in the Years 2004–2006, 27 October 2003. For a discussion of agricultural legislation in Uzbekistan, see Schoeller-Schletter (2008).

prohibition of any form of land sub-lease or other land transaction bypassing the district authorities (article 24).

Until 2003, the transformation of the agricultural system was accomplished in two major phases. In the first phase, the large farms established during the Soviet period were gradually reformed into shirkats, or joint stock companies. In Yangibozor, this was accomplished by 1998–1999. In the second phase, the shirkats were dismantled and their agricultural land was passed to newly established private agricultural enterprises, the *fermer xo'jaliklari*. This process was completed in Yangibozor in 2003 and in Uzbekistan overall during 2004–2006.

From Kolkhoz to Shirkat

During the early 1990s all sovkhozes in Uzbekistan were turned into kolkhozes, because of the government's financial difficulties and the consequent impossibility of its financing the sovkhozes from the state budget. In 1991 the ijara, or leasing contract, system was introduced. This was to become the legal basis for fermers and their enterprises. At first, land belonging to collective state enterprises was made available for lease or rent to applicants, and only small parcels of marginal land were leased, for up to 10 years. Later these conditions were relaxed so that large land parcels could also be leased, for periods of up to 50 years.

Simultaneously, until 1998–1999 kolkhozes were gradually transformed into shirkats. The transformation followed the goal of *sanatsia,* or economic restructuring of the indebted enterprises. Shirkats were compelled to adopt recovery plans to repay the debts contracted by the former kolkhozes, which the shirkats financed gradually selling their non-land assets at auction. In the shirkats, the former kolkhoz members became shareholders (*paychi*) of a joint stock company. But the main transformation that accompanied the shift from kolkhoz to shirkat pertained to the organisation of production.

The initial phase in transforming the organisation of production involved disbanding the brigades, the production collectives by which agriculture was carried out during the Soviet period. These were replaced by agreements between a shirkat manager and the heads of rural households. The latter would work the shirkat's land, which was still owned collectively. These agreements were legally based on waged employment, although, as I describe later, in practice they resulted in a form of sharecropping. The brigades, despite having been officially disbanded, in reality remained a significant administrative and planning force. For instance, tractors still belonged to the shirkat and were centrally housed in motor tractor stations (*motor traktor parki*s), and the use of other agricultural inputs remained planned and executed along brigade boundaries. The heads of the brigades still organised the allocation of tractors and inputs within their jurisdictions, though now the tillers had individual agreements with the shirkat chairman.

Decollectivisation

In the second phase, the shirkats were disbanded and the portion of their land that was sown in state-mandated crops was transferred to fermers. Land that had been held collectively by the shirkats was returned to the state and from there transferred to individual leasers. This process began with the weakest shirkats. Their land was passed to fermers, who were given long-term leases and took over the production of state crops.

As producers, fermers enjoyed greater autonomy than the large state farms had in the past. Nonetheless, their status was in many ways restricted. For one thing, they were bound to a follow-up organisation of the shirkat that in some respects took over the essential functions of steering and monitoring production. In the area covered by my research, that organisation was the *motor traktor parki* (MTP). Elsewhere, fermer and dehqon unions (FDUs) and water user associations (WUAs) played a similar role.

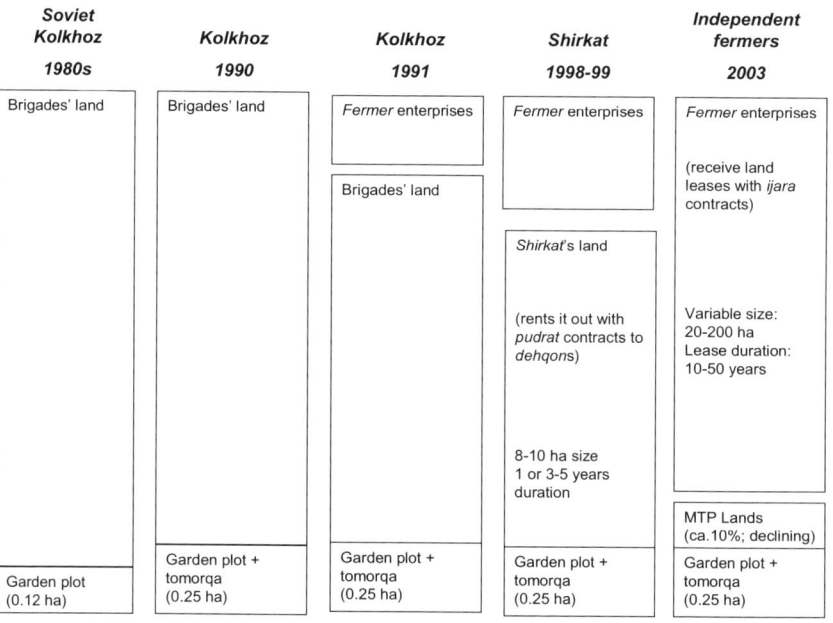

Chart 2. Evolution of land reform in Yangibozor district, 1980s to 2003.

Chart 2 summarises the main steps of agricultural reform in Yangibozor district, showing how the organisation of agriculture transitioned towards full decollectivisation, from kolkhozes and brigades to shirkats, MTPs, pudrat arrangements held by dehqons, and ijara land held by fermers.

Although the law of 1998 introduced and systematised important changes, it would be a mistake to think of it as the clear-cut, watershed moment in Uzbek agriculture. The transition from the Soviet kolkhozes and brigades to the present system took place much more gradually, and different degrees of reform often coexisted for a long time within the same administrative territorial unit.

Like the first phase of the reforms, the second phase unfolded gradually. Fermers began to appear within the framework of the kolkhozes as early as 1991 and then progressively grew in number and gained in share of agricultural land on every large state enterprise. At first they were integrated into the production structure of the kolkhozes, and later they became part of the shirkats within the brigades. Finally they replaced the shirkats, having become dominant in size and number, but still within the framework of an MTP, an FDU, or a WUA, which had by then taken over the steering function that the shirkat once had.

All fermers were gathered into a governmental organisation created to support the development and diffusion of non-collective agricultural enterprises: the Fermer and Dehqon Association (FDA), a national organisation with branches in every region and district of Uzbekistan. Financed partly by membership fees collected from the member fermers, the FDA's priority was to support the fermers; it devoted less attention to dehqons. Its declared tasks included providing help and support in establishing fermer enterprises and in writing contracts with input supply organisations, crop collection points, retail stations, and banks, as well as providing legal counsel to fermers on debt issues in the civil courts and on issues regarding their relationships with state authorities.

In Khorezm, starting in 2000, one or two shirkats in every district were transformed annually into 'unions of fermers'. In 2002–2003, anticipating what would become the future setup of every district, all shirkats in Yangibozor were disbanded, and almost all their agricultural land was transferred to fermers. Administratively, the successor to the shirkat was the MTP. At first, besides keeping the tractors and other machinery once owned by the kolkhoz, the MTP retained a residual, annually decreasing part of the land (about 10 per cent) formerly held by the shirkat. This land remained under direct control of governmental administration, under the pretext of its being a land reserve for future population growth, because the government needed land with which to provide new *tomorqa* plots to the growing number of rural people living on dehqon farms. During my last visit to Khorezm, in September and October 2006, the remaining MTP land had been transferred to newly established fermers, and some of the MTPs in my case study district had been declared bankrupt and closed down, thus concluding the transition from shirkats to fermers.

After decollectivisation, a third phase, that of farm consolidation, was carried out nation-wide in late 2008 and early 2009.[117] Small and medium-size fermers, whose leased allotments did not exceed 80 hectares, were merged into a prescribed number of brigade-size farms in each former shirkat. The farm consolidation policy considerably reduced the number of fermer enterprises in each former kolkhoz-MTP by augmenting the land holdings of the remaining fermer enterprises. The land leasing (*ijara* contract) principle was maintained. A further trend seems to be that of replacing the system of state–ordered crops and procurement prices with a new form of taxation.[118]

New Regulations and Land Use Patterns for *Dehqon*s and *Fermer*s

The two key actors in Uzbek agriculture today are dehqons, or peasants, and fermers. The characteristics and legal statuses of both are described in the land code of 1998. Dehqons are members of rural households that in the past were affiliated with a kolkhoz, who served the kolkhoz as workers or other employees such as teachers and administrators. Starting in 1989–1990, such households were each entitled to a subsidiary plot (*qo'shimcha tomorqa*) of 0.13 hectare of irrigated agricultural land, in addition to the 0.12 hectare of land that was attached to the house during the kolkhoz period as a garden for private use.[119]

---

[117] The farm consolidation policy was initiated on the basis of the presidential decree 'On the creation of special commissions for farm size optimisation', P-3077, 6 October 2008.

[118] Telling in this respect are attempts to calculate the indirect taxation of cotton producers and the recommendation for introducing a higher land tax while liberalising the domestic price of cotton (see Djalalov 2007: 98).

[119] In the past, kolkhozes and shirkats were the bodies entitled to allocate tomorqa land to households that requested it. With their demise, this duty passed to the district land measurement office, which has delegates in the MTPs (the shirkat-level land measurers). During an interview, a land measurer explained to me how one received a tomorqa: 'One writes a letter of request [*ariza*] to the head of the village council in which he applies because his family is too big to live off of one tomorqa. The head of the village makes a decision and informs the land measurer. There are two variants: the first is that 30, 40, or 50 families take a large plot [*kontor*] split into single parts. The other variant is that if plots are available close to the village [*posolka*], they can give that piece'. Dehqons preferred the latter version, because otherwise plots might be located at a great distance from their homes.

Dehqon families today represent the majority of Uzbekistan's rural population. Their production is mainly for subsistence and is entirely household based. The basis for the livelihoods of these 'minifundia' (Griffin, Khan, and Ickowitz 2002) is a subsidiary household plot (0.13 hectare), a small garden plot, and a shed where people keep privately owned livestock. In irrigated areas, the land, by law, may not altogether exceed 0.25 hectare.[120]

In 2004, dehqon families supplemented their incomes through work on a shirkat or with a fermer. These were family-based contract arrangements (*oilaviy pudrat*) to look after a parcel of land sown with the state-commanded crops, cotton and wheat.

Unlike dehqon farms, fermer enterprises are legal entities established by resolution of the district hokim with certain requirements for reporting their financial operations and transactions. Beyond their legal status, dehqons and fermers differ in the types of land contract arrangements they have with the authorities. Dehqons have a rental arrangement (*pudrat,* from the Russian word for 'to rent'), and fermers have long-term, contract-bound leases, or ijara.

As early as the kolkhoz period, family-based brigades were allotted parcels of land sown in state crops to look after, as a measure to enhance productivity. This arrangement continued in the shirkats in the form of pudrat arrangements, in which former brigade workers entered into contracts with the shirkat chairperson to allocate them specific plots of land. The contract could be annual or could extend from three to five years. In Khorezm, the average size of a plot covered by a pudrat arrangement was 8 to 10 hectares. The size ultimately depended on the capabilities of the dehqon family; the land allotment would be adjusted according to the size of the family workforce. Local availability of land was another factor. On pudrat land, the head of a dehqon family was employed by the shirkat to grow the crop. A pudrat head was also appointed to apportion the plot if different families were involved. The shirkat provided seeds, fertilisers, and other necessary inputs.

Pudrat work was to be paid for in a monthly wage, but because of cash shortages in the *rayon,* salaries were usually paid in kind, in shops that, during my fieldwork in 2003–2004, were still affiliated with the state retail system. Prices at these shops were higher than those at the local bazaar, which depreciated the value of the salary. In theory a worker's monthly

---

[120] In Khorezm, because of land shortages and high population density, the size of the small-plot of dehqon farms has been further reduced in recent years. The 0.25 hectare of land has come to include the house, the cowshed, roads, and so forth, so that the area effectively available to households for cultivation in many cases amounts to 0.19 hectare.

pudrat salary was equivalent to about US$10, but this amount did not reflect actual earnings, because expenditures for things such as electricity, pension contributions, and communal services to the household were subtracted. What remained, made available in kind through the shops, had a market value of some US$5–6 monthly.

Because wages earned through pudrat contracts were insufficient to cover families' living expenses, pudrat in fact became a sort of informal sharecropping system. By agreement with the shirkat rais and brigadier, families were compensated for their work on shirkat land with a share of any harvest in excess of the state procurement quota. This, along with the possibility of using shirkat land for a second yield, with no delivery to the state, greatly increased the attractiveness of the pudrat contract for the tillers. They carried out all the work in the fields while the shirkat provided inputs, tractors, and cover for what was, strictly speaking, an illegal use of collective land. The rais and brigadier also kept a significant share of the harvest.

Although pudrat work, strictly construed, referred to employment on a shirkat, the term was also used colloquially to refer to work on fermer enterprises, if workers were formally employed and their workdays were signed off on an employment card. In 2004, most pudrat work by dehqons in Yangibozor was actually work for fermer enterprises.

Fermers, by contrast, obtained their land by ijara contract, a long-term lease on land that was bound to the cultivation of certain crops as specified in the fermer's business plan. Under this plan, the titular farm owner applied for land at the farm establishment commission (*fermer xo'jaliklarni tashkil qilish kommissiyasi*), the shirkat, or, once the shirkats were dismantled, the office of the district hokim. Applicants for land leases and for the title of fermer had to specify in a business plan their cropping schemes for the next 10 seasons. Changes in the cropping scheme required authorisation from the agricultural branch of the district administration (the RaySelVodKhoz, known colloquially as the *boshqarma,* the former Agroprom).

Although in theory any citizen could apply to become a fermer, in practice a fermer was usually a dehqon who had become the owner of a fermer enterprise after obtaining a long-term land lease from a shirkat or, later, from the district governor. The duration of the lease depended on at what stage in the reforms the fermer enterprise was established. In the early stages, renewable leases of 10 years were given; later, leases were extended up to 50 years.[121]

---

[121] Lease duration was first of all dependent on the terms of use determined by the state. With the farm consolidation policy of 2008–2009, the majority of fermers lost their ijara leases before their formal expiry. The internet newspaper *Delovaya Nedelya* reported that fermers in the Tashkent region with holdings smaller than 80 hectares were asked to sign an agreement

According to law, fermer enterprises signed yearly contracts with state retail stations – the cotton ginnery or the wheat processor – on the basis of which they obtained credit for government-subsidised inputs (fertilisers, seeds, tractors, etc.). Input credit, however, did not pass through the farm accounts but went directly to input suppliers, so farms and fermers were only indirectly involved.[122] Farm profits were calculated after the harvest, and farm expenditures were subtracted from the farm's bank accounts on the basis of the amount of harvest delivered to the retail station. All prices were set by the government and were valid over the entire republic. Other trade in state crops and subsidised inputs was prohibited. In order to support their development, newly established farms enjoyed a tax holiday of two years.

According to their (national standard) contracts, fermers were asked to reach production targets of a certain amount of cotton per hectare. This target was calculated on the basis of the quality of their land plots, expressed through a classification system called the *ball-bonitet*, a holdover from the practices of the kolkhozes and sovkhozes.[123] During my fieldwork, the average output to be reached by fermers in Khorezm was 2.9 tonnes of cotton per hectare. In theory, fermers were free to sell the cotton they harvested in excess of their target amount. In practice, at the time of my fieldwork, licences for trade and export of cotton were not granted in Khorezm, which resulted in the state's holding a monopoly on cotton.

Cotton pricing was centrally regulated and depended on the quality of the delivered cotton, which in Yangibozor was determined in the laboratory of the cotton ginnery upon delivery. There, cotton was supposed to be classified according to five quality levels, which corresponded to different price categories. In practice, the determining of cotton quality was never very accurate. Prices ranged from 36,355 to 146,813 *so'm* per tonne of raw cotton in 2003, and from 65,460 to 263,890 so'm per tonne in 2004. The quality and therefore the price of the delivered cotton diminished over the harvest season, September to November. In 2006 a fermer told me that the price per

---

of voluntary abandonment of their leased plots. A. Saidov, 'Time for Latifundia: Bad harvest induced the government of Uzbekistan to enlarge the farms', *Delovaya Nedelya*, 15 November 2008 (www.dn.kz/index.html).

[122] In 2004, input credit was issued at an interest rate of 5 per cent for cotton and wheat.

[123] According to Kienzler (2010: 6), the *bonitet*, or 'soil bonitation', is 'a classification system for soil fertility ranking land quality of particular soils on a 100-point scale depending on parameters such as groundwater salinity levels, soil organic matter ... and gypsum content in the soil'. Soil 'bonity' maps were created for all collective enterprises during the Soviet period. They were still used during my fieldwork in Khorezm to determine production targets for cotton and wheat on every *kontor* and to calculate the amount of fertiliser to be used on that plot. For instance, in the average case of a soil with 60 points of 'bonity', the target cotton harvest was to be 2.9 tonnes per hectare, and agricultural inputs were calculated accordingly.

kilogram paid by the cotton ginnery in Yangizobor had grown to 358 so'm. In this conversation it emerged that cotton's profitability and attractiveness for fermers were growing in comparison with previous years. According to fermer M., in 2006 fermers reached average net earnings of 1 million so'm (about US$800 at the time) for every three hectares grown in cotton.

Winter wheat production, although state directed, entailed greater freedom for producers than cotton production. The average winter wheat target of 1.8 tonnes per hectare (varying according to the land's *ball-bonitet* value) left an ample share of the total harvest at the fermers' disposal, at least in theory (harvests averaged five to six tonnes per hectare in Yangibozor in 2004). Furthermore, in Khorezm the early harvesting of winter wheat allowed for a second yield over the summer, whether of vegetables, fodder crops, or, if sufficient water was available, rice.

Land Use Practices for Freely Marketable Crops

For Khorezmian producers, the comparatively low prices paid by the state for cotton and wheat made freely marketable crops such as rice, vegetables, and watermelons extremely attractive. The state authorities restricted their cultivation, however, to a limited share of the total available irrigated land.

Sharecropping agreements between the rais of a kolkhoz and kolkhoz workers regarding crops other than cotton had already begun to be made during the last years of the Soviet Union. Similar to the way kolkhoz workers could make *arenda* arrangements to lease small plots of land (Visser 2008: 36), during the shirkat period a dehqon could contract a *tender*, an agreement to grow freely marketable crops on collective land. *Tender* agreements (usually applicable to one to five hectares) were officially introduced in the early 1990s in reaction to the fall of productivity caused by the unattractive remuneration for state crops. Referring to the mid-1990s, a former rais I interviewed in Shovot district explained that he felt obliged to devote a part of the land of his shirkat to *tender* arrangements in order to keep enough workers motivated for the compulsory cultivation of cotton. In the *tender,* a dehqon could get a plot of land on which to grow crops either on a sharecropping basis with the shirkat or for an annual cash payment to the shirkat. Unlike an ijara lease, the agreement lasted a maximum of four years, and the dehqon did not need to legally establish a farm.

After the dissolution of the shirkats, *tender* agreements gradually came to an end. In the MTPs Bo'ston and Madaniyat, some *tender* agreements still existed on the MTPs' land reserves, but new *tender*s were not being granted, and it seemed that after the leases expired they would not be renewed.

With decollectivisation, the attractiveness of growing freely marketable crops remained undiminished. Although land transactions that bypassed the authorities were illegal, the practice of 'buying' land, through sharecropping agreements or in exchange for cash, on which to grow crops that were freely marketable in the bazaars or consumable for households continued. Apparently, a significant amount of land in every former shirkat was cultivated under informal agreements. In such unregistered deals, the profit from the land was immediately available to the contractors, without the mediation of the state.

This black market in land encompassed two distinct types of activities, one more subsistence oriented and the more entrepreneurial. The first form was generally tolerated by the district authorities. Like the shareholders in the shirkats, fermers lacked cash to pay their workers, so they granted them usufruct in plots of land instead. These plots were easily recognisable because they were small and usually situated at the edges of large fields grown in state crops. Marat, son of a former rais, owned a large farm, 34 hectares, which he had established during the privatisation of a shirkat. In 2004 he explained to me that he had to make such informal arrangements with the labourers on his farm out of his sense of responsibility towards them:

> There are 12 families that must be fed out of this land. Of these, at least 26 people come to work here. I have 23 hectares of cropped land, of which 22 hectares are cropped under the state plan. The bank does not give us our wages in cash. What can I do? I have to care about them. For instance, one of them I have sent to Kazakhstan to work for cash. On my farm I had some hectares in an orchard, but unofficially [*norasmiy*], not on the maps. Two hectares of the orchard I have now prepared for rice. I have no cash to pay the workers, so I will give them a bit of this land. Every family has a bit of such kind of land. The profit from this kind of land remains all in the people's hands.

As I show in the next chapter, in the context of the relationship between fermers and dehqons, the expression 'to give land', which Marat used, often entails a sharecropping arrangement.

Besides entering into unofficial sharecropping arrangements, people might sell their entitlements to land unofficially, for various reasons. For instance, a household that received a tomorqa plot situated too far from the family house might try to make a deal for money or swap its plot with that of another household. In a village in Bog'ot district, for instance, a dehqon was apparently trying to sell the use of his tomorqa for a rice season for 15,000 sum, because no one in his household could take care of it during that time.

Altogether different from such small-scale, informal arrangements among dehqons or between dehqons and fermers is a more speculative type of unofficial land use. Usually it regards larger plots, of many hectares, that large fermers (and formerly, shirkat officials) sub-lease to capable agricultural entrepreneurs for cash or under a sharecropping arrangement. This is riskier than informal deals involving small plots, because the control of law enforcement agencies is tight, and penalties are high. I discuss the importance of such arrangements in chapter 6.

Privatisation

Before the end of the shirkats, in Yangibozor as elsewhere in the country, the gradual privatisation of enterprises, buildings, and assets formerly owned either collectively or by the state paved the way for full decollectivisation. In Yangibozor, small shops were privatised immediately after independence. In 1992 and 1993 all houses and other privately used buildings, owned in the past by the state or the kolkhozes, were privatised next. In 1994 and 1995 some state-owned construction enterprises were restructured into holdings with mixed private and public ownership, usually with the number of state-controlled shares decreasing over the years. Alternatively, such enterprises were straightforwardly privatised in public auctions. A particularly important year for privatisation in the district was 1998. Concomitant with the introduction of the shirkats, the district cotton ginnery, established in 1969 and equipped with machinery brought from a manufacturing plant in Tashkent in 1987, was legally reformed into a shareholding, and a significant number of its shares were sold by the state to private entrepreneurs. In 2004 a former district deputy hokim privately owned the largest stake in the company.

In the years following the introduction of the shirkats, privatisation proceeded with the selling of shirkat assets – tractors, machinery, and buildings – in order to accomplish the economic 'sanitation' (*sanatsia*) of the shirkats. Unlike land grown in state crops, shirkat land devoted to orchards could be sold at auction, because although the land remained state property, the trees did not. Formally, only the trees were auctioned, but the value of orchard land at auction was defined by the size of the plot rather than by the number or condition of the trees on it. Privatised orchards were greatly desired, because they could be used to grow not only fruit trees but also freely marketable crops that were exempted from state procurement quotas. This first wave of land 'privatisation' – really the selling of entitlements to usufruct of orchard land – proved to be very important for rural residents, because every shirkat had part of its land in orchards. Orchard plots were apportioned in small parcels, often of just one hectare, in order to spread them among a large number of dehqon households.

These auctions were planned and held locally. Preceded by the preparation of an inventory by the staff of the shirkat, the auctions were carried out by commissions installed by the district government and, in principle, according to the rules of a statute defined by the ministry of agriculture. In practice, ample margin for discretion was left to the local heads of the shirkats, who abused their positions to take advantage of the privatisation, such that it became, in the perception of those who were excluded, 'theft' (cf. Allina-Pisano 2008: 190; Nazpary 2002: 34). According to R., a fermer who came away empty-handed from the orchard auction in his village, privatisation had been highly unfair ('The auction has been unjust [*auktsiya noto'g'ri bo'ldi*]'). Although the number, condition, and sizes of the trees on the plots differed, the plots had been sold according to their measurements. Higher prices had been demanded for large, half-empty plots than for small plots with good trees, which were taken by the shirkat bosses.

In Yangibozor, commissions for privatisation and the founding of farms, which were installed ad hoc in every shirkat, carried out the final step of decollectivisation. During this phase, beginning on 1 January 2003, all residual collective land was transferred to the newly established private farm enterprises.[124] According to law, the commissions, each composed of a large number of state officials and representatives of government organisations, were to rule autonomously on applications for farmland, with successful applicants becoming fermers. Selection was based on candidates' suitability as determined by an attitudinal test, educational background, and proven capacity to manage a farm.[125] The work of the commissions consisted in defining the plots on which fermer enterprises were to be created and selecting capable applicants.[126] As many as five applicants competed for a specific lot of land. The commission members graded the candidates on their agricultural and entrepreneurial skills, and the candidate at the top of the list received the lease and the right to establish a farm.

Although the introduction of the commissions was meant to ensure that land was conferred on the most skilful and capable individuals, it still could not prevent land from being treated as an informal commodity. The shirkat administration considered all applications and selected those whose farm establishments were to be sustained. The decision depended, finally, on the district hokim's ratification, which was rendered official by his releasing

---

[124] A decreasing reserve of 5–10 per cent of the former shirkat land was not allocated to fermers but was placed under the accountancy of the MTPs.

[125] O'tamov and Qurbonov (2001) described the legal modus operandi of the commissions for a district near Yangibozor.

[126] See map 3, in chapter 2, which was created by the commission for the disbandment of the shirkat of Bo'ston.

the farm establishment decree (*qaror*). At this point, the head of the private farm signed the lease (*shartnoma*) with the shirkat rais. Finally, the land measurer (*zemlemer* or *yer tuzuvchi*) measured the plot and recorded it in the land register. The leased land had to be planted in the crops specified in the fermer's business plan. This was the state's not-so-concealed strategy for maintaining its grip on cotton and other crops: a business plan 'suggested' by the authorities enhanced one's chances of becoming a fermer.

The law regulated farm establishment and land lease procedures, but in the perception of some fermers I interviewed in Yangibozor, land allocation, behind the screen of the official procedure, followed the flow of money. Like the privatisation campaigns that preceded it, land allocation ended up becoming a cash machine in the hands of the local command hierarchy. The actual procedure for getting farmland, I was told, was as follows: The prospective head of the farm wrote a letter of application to the shirkat rais and the *rayon* hokim, requesting a specific plot. Previously, he had negotiated with the rais to agree on the plot to be formally requested. Although it was seldom admitted, this procedure was often accompanied by a bribe paid by the applicant to the rais. Unofficial prices for land differed according on the quality of the land. According to Eckert and Elwert (2000: 22), the 'privatistic' approach of the elites

> leads to the kolkhoz being the sinecure of the chairmen, their families and the administration. This means that kolkhoz assets are used to pay personal bribes; that kolkhoz land is distributed to relatives and friends; that kolkhoz contracts are made for the personal profit of the administration or parts of it. The farmers become the Golden Goose, without the chairmen realising that they actually inhibit an increase in production by depriving the farmers of incentive and of capital.

As much as the kolkhoz elites saw fermers as targets for rent extraction (the 'golden goose' of the fairy tale recorded by the Brothers Grimm [Brüder Grimm 1812]), so do the *raislar* since decollectivisation, but they have adapted their attitude to the contingencies of the new situation. In the account of a friend of mine, the end of the shirkat and the privatisation of its land and assets just meant more business for those who had the power to define the resource flows:

> F: In my village, everyone who establishes a fermer enterprise has to pay a bribe, let's say 150,000 to 200,000 so'm per hectare. Whoever offers more gets the land. This means you pay two to three million so'm and the rais of the shirkat takes this money. Out of a thousand families, only 40 can become fermers. There is competition. This means that the rais can make a lot of money.

TT: What about the hokim?

F: Applicants for farmland do not talk with the hokim, but the hokim talks with the rais, afterwards. It will be like this: 'Rais, I gave you 1,000 hectares of land. Now give me my money'. At this point the rais keeps a quarter of the money for himself and passes, say, 1.5 million so'm out of 2 million to the hokim. The hokim has invested much money to get his position. He also has to give to his superiors; if not, he will lose his job.

Although the figures given in this example were hypothetical, the logic behind it was confirmed in several interviews. Thus, behind the official façade of a complicated procedure, in Yangibozor decollectivisation was meticulously prepared for in advance.

For the authorities of the district, the problem with implementing the decree on decollectivisation was that applicants who were economically capable of taking on the burden of a farm-founding contract and experienced enough to run a farm were scarce in the villages. The new farmers inherited the debts of the shirkats, so they started their operations with large deficits. In order to spread the debts among many households that otherwise would have been incapable of carrying the financial risk entailed by a large farm, more farms, reduced in size, were established than had been called for in the hokims' original plans. Farm size basically was adapted to the capabilities (and sometimes to the exigencies) of the designated applicants. In the following account, quoted from my field notes, Oktyabr (who preferred to be called by the more Uzbek-style name O'ktam), an Uzbek geography teacher at a village primary school and an *elatkom* officer, explains why he failed to become a fermer:

O'ktam works as a *pudratchi* with his wife, his son, and his daughter-in-law on the newly established farm number 3. At the time of the land distribution planning, he was interested in becoming a fermer on lot number 7. The commission was led by A.J., at that time an important exponent of the district hokimiyat, and today chief agronomist in the MTP of the neighbouring village. A.J. asked the shirkat staff, 'Who is there [*kim bor*]?' They said that there was O'ktam, and so A.J. asked him whether he wanted to become a fermer. He did, but the problem was that one had to have the economic capabilities to run a farm, and O'ktam did not, so it was not possible. O'ktam gave up, and after three days another candidate was found. O'ktam explains that agriculture is very difficult to manage. 'You have to invest much money, you have to take risks'. For him, a farm of three to four, a maximum of five hectares would be feasible and appropriate for the small workforce of his family. 'We do not have tractors, we are not a large family, there is the problem with water shortage. If you don't have capital you will run into troubles. And if you take labourers, you have to be able to pay them on time, otherwise they will not work and you will go bankrupt and lose everything. Twenty hectares were too much for me. I had to say no'.

Contrary to the assumption that passing the shirkats' land to fermers would start a run on land, in most cases people were afraid of the risks of farming. The commission head's question, 'Who is there?' is significant because it reflected an attempt to identify the elements that made the old system work – that is, the extended families who managed the brigades – and transpose them into the new system. This did not always succeed. The old elements, as O'ktam's case tells us, were not always suited to the new criteria called for in decollectivised agriculture, including capital and other resources such as ownership of agricultural assets, a predisposition towards risk, and workforce availability.

Although the façade of the procedure was one of meritocracy, with education, experience, family size, and capability the criteria for applicant selection, in the end those who had held high-ranking positions in agriculture before decollectivisation got to lease the largest and best pieces of land. Local-level reforms privileged former kolkhoz members who had the means to pay bribes for fertile, well-irrigated land. Other people had to be convinced to assume the burden of farming the remaining, unfertile and saline land plots that nobody wanted, and they were sometimes even pushed to do so. Most of the newly established fermers received land under unprofitable leasing conditions, which turned the lease into a burden – in Verdery's (2004: 156) famous phrasing, it turned some 'goods' into 'bads'. As in other decollectivisation contexts (Allina-Pisano 2008; Verdery 2003), land disputes, competition over land and its allocation, conflicting claims, and conflicting ways of legitimising those claims accompanied the process.

One case, which first aroused my curiosity because of the unusual name of the farm involved, is a particularly bold but otherwise unexceptional example of the turbulence of decollectivisation. The farm, part of the former Xalqobod shirkat in Yangibozor district, was somewhat pretentiously called 'Qushbegi', a title for a high official once used in the court of the khan of Khiva; it was the equivalent, a villager told me, of 'prime minister'.[127] Al-

---

[127] According to Bregel (1986: 273–274), under the Qongrat dynasty (eighteenth–nineteenth century) the *qosh-begi* were among the highest officials in the khanate of Khiva: 'The qosh-begi always belonged to the Uzbek nobility, the *amirs* (and sometimes was a relative of the khan), and was in charge, mainly, of military affairs. Besides that, the qosh-begi governed the northern part of the khanate of Khiva inhabited by nomadic and semi-nomadic Uzbeks, Karakalpaks and Turkmens (that is, he mainly supervised the collection of taxes in this region)'.

though the farm was registered in the name of Oybek Tojimov, everyone in Xalqobod knew that Qushbegi – by far the largest fermer enterprise in the former shirkat – was owned by Otanazar Ahmedov, the former rais of the shirkat. Ahmedov started the farm in 1999 with 10 hectares and in January 2003 increased it by a further 105.7 hectares. His land was excellent, well suited for irrigation, and had many trees.

Ahmedov, a man in his sixties at the time of my fieldwork, was a grandson of the last *qushbegi* of Khiva, whom the Bolsheviks exiled to Siberia. His father, 'a rough and simple person', had been a brigadier in a kolkhoz in a different area. In January 2001 Ahmedov was made rais of Xalqobod, in order to carry out the privatisation of the collective's assets and prepare for its dismantling. Before that he had worked, among other places, in the rice-growing shirkat Jayhun. During privatisation, he managed to move tractors and other machinery and valuable assets from the shirkat to the account of his own farm.

But the land that Ahmedov managed to get for himself had been looked after for years by another person, No'rmetov Bekmat, who was known by the nickname *daxma* (gravestone). From the beginning, Bekmat daxma's sons raised claims to the land that had long and unquestionably 'belonged' to their father when he was brigadier. They argued that since their father had worked on the land for many years and enhanced its quality, they had a legitimate claim to receiving it upon the dissolution of the shirkat. Before the land of their father's brigade was turned into the lot that Ahmedov eventually obtained, they even built on it a *dala-shiypon,* a building used in the summer for work breaks and for storage, in order to strengthen their claim.

The rais, however, as a member of the land commission for the shirkat, had managed, during the planning for privatisation, to define the lot he was interested at such a large size that no one else in Xalqobod had enough 'strength' (*kuch*) to win the bid for it. Bekmat daxma's first son, Bektosh, publicly protested against Ahmedov for what he perceived to be an illegitimate land expropriation, but without result. Unable to regain control over the land, Bektosh even went to ask the hokim for justice, but the hokim said that the farm establishment decree had already been published, so nothing could be done. Bektosh, son of a once powerful and well-respected brigadier, remained landless and in 2004 worked as a butcher in the village.

The 'rise' of Otanazar Ahmedov, descendant of a bey, was atypical among the fermers I interviewed. Soviet rule brought sharp discontinuity in the reproduction of the elites, and people affiliated with former bey landlords were prevented from accessing prestigious jobs or education. In a conversation about the social origin of today's hokims, a fermer and son of a rais told me, 'Those who were "high" at that time [before the Soviets] could not work in top positions. Not even in our district. You could not become rais if your father was a bey. The KGB controlled [everything], and it was a perfectly efficient machine'. A new elite, recruited from the former lower classes, was installed and grew with the kolkhozes, and the transmission pattern remains unbroken after decollectivisation. During my fieldwork I came across five cases in which the father of the rais of an MTP had been rais of a kolkhoz before him. Not all chairmen 'inherited' their positions, but in Khorezm this was not infrequent. Decollectivisation has been an important moment for the social reproduction of the elites that were already in power during the kolkhoz period. Nowadays in Khorezm, however, 'being able to claim descent from the Khivan khans or from any other high-ranking officer in the former khanate seems to be a source of prestige' (Kehl-Bodrogi 2008: 48). Such claims have multiplied, although most of them are false.

As seen in the former rais Otanazar Ahmedov's success in obtaining a large farm and in the unsuccessful case of the teacher O'ktam – cases no different from many other postsocialist scenarios (Verdery 2003: 93) – decollectivisation privileged the former elites of the kolkhoz. Those who in the past had the means and the power to accumulate resources now saw their chance to emerge as successful fermers. In this way, Uzbek land reform stimulated the return to agriculture of capital and resources that had earlier been illegally withdrawn from the kolkhozes.

Ahmedov's and O'ktam's stories also show how the district authorities, while deciding, in their top-down planning and implementing of decollectivisation, who would and would not receive land, stumbled upon sensitive 'boundaries' drawn locally by kinship relations and how they attempted to mediate between plan desiderata and family claims.

Decollectivisation in Yangibozor

Yangibozor was the first district in Khorezm in which agricultural land was entirely passed on to fermers. In 2001, 364 fermer enterprises were already established within the 11 shirkats of the district, but the large bulk of such enterprises were established in the winter of 2002–2003, upon the dissolution of all shirkats.[128] By 2006, decollectivisation was entirely accomplished. The number of fermer enterprises in the district then reached 1,406. The

---

[128] By 1 January 2003, when the 11 shirkats of the district were dismantled, 1,164 fermers had obtained most of their arable land, together with the labour of some of the former kolkhoz workers who had been linked to the shirkats through pudrat contracts. In the first year of decollectivisation the 11 successor MTPs maintained a remaining 8 per cent (1,339.5 hectares) of former shirkat land (each got 5–10 per cent of the land formerly managed by their shirkat), which they continued to manage as these reserves diminished annually until 2006.

enterprises established after 2003 were introduced gradually on the residual land reserves managed temporarily by the MTPs.

Fermers, although equal in juridical status, were far from being a homogeneous group. They differed widely in background, specialisation, and size of farm. Table 1 shows the variation among private farms in Yangibozor in 2004 by size and crop specialisation. By far the most important category, in terms of both number of farms and hectares planted, is cotton and grain farms, officially called *dehqonchilik* farms. They deal with the core business of cotton, wheat, and rice, the crops that matter most to the district authorities and to the governmental budget, and therefore they face considerable control and interference by state authorities.

Table 1. Farm specialisation in Yangibozor district, April 2004.

Type of Farm	Number of Farms	Total Hectarage	Average Hectarage
Cotton and grain (wheat, rice)	713	17,426.1	24.40
Orchard	331	645.6	1.95
Grapes	31	44.9	1.40
Vegetables	26	87.9	3.30
Cattle	18	319.6	17.8
Silk	31	62.4	2.00
Fish	5	69.3	13.86
All farms	1,164	18,656.8	16.02
MTP land	10	1,339.5*	133.95

Source: Author's fieldwork data, 2004, based on data from the Yangibozor district land measurement office.
*Of which 12.5 hectares was in orchards.

Orchards made up the second largest number of small fermer enterprises in 2004. Ranging between one and four hectares, they were less dependent on the state-controlled input and retail structure. Instead, their production was oriented towards the needs of the owning household. Often, the fermers who got these plots at auction had worked in the orchards before, or they were retired employees of the shirkat with distinguished careers. Orchard farms were attractive because they were exempt from state command. Ordinary people appreciated them because they enabled the owners to earn modest livings relatively unimpeded by bureaucracy. Because of these peculiarities, orchards should be considered a category apart. 'State crop' farms dealing with wheat and cotton had to be larger than orchards in order to meet the requirements of the state agroindustry.

Cattle and vegetable farms were incorporated into the state-ordered procurement system differently from farms specialised in cotton and grains. Like cotton growers, cattle raisers had to sign contracts to deliver meat to state-controlled retail enterprises and bazaars. Their profitability depended on the terms of their contracts, which could make cattle raising less attractive than it might have been otherwise.

In Yangibozor in 2004, most small private farms, in the range of one to four hectares, were orchards (table 1). Farms of this size were more amenable to the small-scale farming of dehqons than to cotton-growing by fermers, who had taken over the burden of the state plan from the shirkats.

Among the fermers dealing in state crops, the smaller farms (10–30 hectares) growing mostly cotton and some winter wheat tended to be economically vulnerable. Their owners' living conditions were not far above average for the rural population. Larger fermers, with farms ranging from 40 to more than 100 hectares, belonged to the class of well-off rural notables, economically capable and with privileged social backgrounds. These fermers were often the same people who formerly played important roles such as those of rais and executive staff member in the kolkhozes.

The government's vision for the agricultural sector assigned fermers and dehqons different roles. Fermers held the reins of production in both a positive sense (capitalising on economic opportunities) and a negative one (being responsible for accomplishing the plan and liable for failure). Dehqons played the role of suppliers of (cheap) labour. Dehqon families were supposed to live on subsistence agriculture but also to develop in other sectors. During my fieldwork, the wealthiest villages in Khorezm were those in which the rural population relied on crafts and trade, in addition to agriculture, as sources of income.

The government's consolidation policy of 2008–2009 accentuated this division. The perspective that farming has been transformed into a competitive undertaking, in which the less successful will drop out and those remaining will expand at their expense, has now reached the minds of the agricultural actors in Yangibozor district. The scenario in which fewer but larger, more economically capable farms, with increased productivity and greater capacity to invest in agriculture, have taken over the ijara tenures of others, has been realised more quickly than expected by most fermers who started their enterprises in 2003–2004.

'Fairness' and New Rules of Agriculture

In the rhetoric of the government's agricultural reform policy, the new rules of the game are presented as being equal and fair. The role of the authorities is purportedly to support the reform process by helping fermers stay on track. Yet the 'new-but-old' input suppliers and output marketing organisations on which the fermers depend are still controlled by the district authorities, who use their power to influence and manipulate decisions. At the same time, the profitability (or otherwise) of farming depends heavily on the specific arrangements fermers are able to make with the web of input and output organisations. Chart 3 shows that many of the operations fermers have to carry out for their enterprises are actually managed entirely by external institutions such as the MTPs and the input supply organisations, which are under the control of the district authorities. Transactions such as those between the bank and the cotton ginnery or the wheat collection point skip fermers' participation altogether. Through these new-but-old organisations, the district authorities control production just as they did in the past, even if on paper the organisations have been privatised.

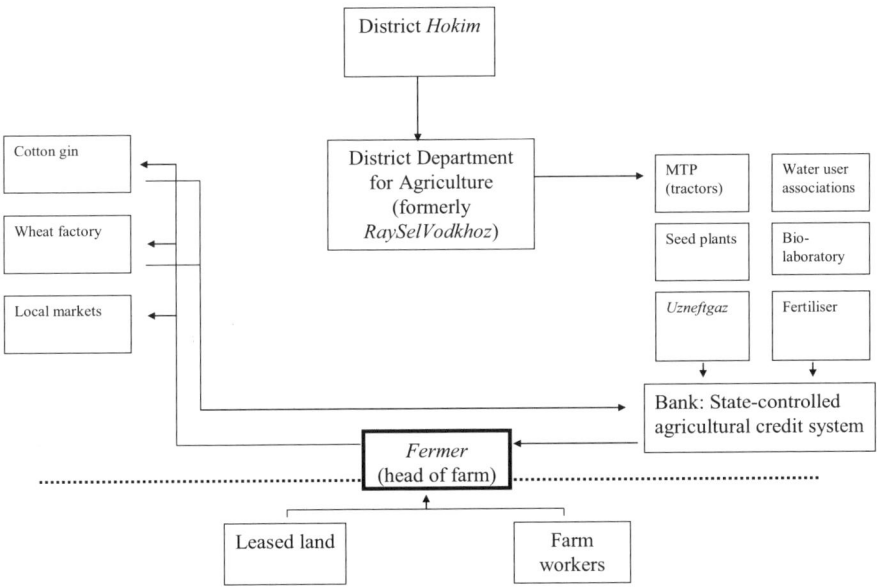

Chart 3. The organisation of farm production in Khorezm as of 2004.

In Yangibozor, tractors are the scarcest and most needed goods during all phases of the agricultural cycle for state crops. Most private farmers depend on the services offered, at high cost, by the local MTP. A minority of fermers own tractors, but even those who do are urged to use the expensive maintenance services provided by the MTP. Legitimated in its role as input provider and supporter, the MTP in reality restrains the autonomy of private farmers, first because of its quasi-monopoly over tractors and machinery, which re-creates, in the footprints of the shirkat, relations of dependency between producers (farmers) and the executive staff of the former collective. The MTP offers the services of its aged tractors at a much higher cost than the competing services of privately owned tractors, and the leasing or renting of unregistered tractors is illegal. Farmers are held to the letter of the law by their obligation to account for every economic transaction, which makes it difficult for them to lease or rent unregistered tractors. This structure is reinforced by the fact that because of high customs duties, no new tractors are being imported, which effectively keeps the system closed.[129]

In my conversations with fermers, they repeatedly highlighted the importance of owning a tractor. For example, when I asked one fermer, 'What do the fermers lack to make money?' he replied, 'Tractors. For eight hectares of rice you will pay 400,000 to prepare a field once. If I had a tractor, just 200,000. And you need the tractor again and again, so you see, the tractor makes the big difference'. According to M., a fermer from Xorazm MTP, the selling of shirkat assets not only had been unfair but also rested on a wrong interpretation of the law. Because the collective enterprises, before decollectivisation, were owned by their members and not by the state, the establishment of fermer enterprises should have entailed the privatisation of tractors and other collectively owned assets. But the state kept the tractors, and now all fermers faced large debts to the MTPs.

Official rules required farmers to rely on the state-managed system of input supplies, yet its shortcomings forced them into illegal arrangements for their economic survival. This situation was not limited to tractors but extended throughout the whole range of agricultural inputs. In this respect, a special role was played by the reforming banking system, the shortcomings of which had also been a drag on the development of the private farming sector (Xolliev 2003).

---

[129] At the time of my fieldwork, the Uzbek-American joint venture CASE was an exception. This company, operating throughout the country, had introduced a limited number of new tractors, which the government had accepted because of its extreme shortage of tractors. With its monopoly position, however, the company had not affected the pricing of the tractor services market for farmers.

The district's tight control over agriculture was exemplified by the cropping arrangements on which fermer enterprises had to agree with the district authorities. In Yangibozor, growing rice could be 10 times more remunerative than growing cotton, but growing rice was a risky business if the district authorities did not release permission to receive inputs and market access. Nevertheless, fermers were tempted to grow more freely marketable crops such as rice, at the expense of state-ordered cotton, in the hope of maximizing their profits or because they needed to earn cash to pay their labourers or obtain tractor services.

The shortcomings of the regulatory system thus drove many fermers into a situation of near-illegality or para-legality, which in many respects resembled the 'suspended punishment' of the Soviet system in the past (see Rasanayagam 2003). The authorities fostered this situation because it rendered fermers vulnerable. They could be blackmailed and forced to agree to burdensome state quotas or to make donations to assure their survival politically. Because of this precarious situation and their dependence on a weak supply system, fermers knew that their future existence depended entirely on the goodwill of their district hokim, even if on paper the leases for their farms lasted for many decades.

Much as in the Ukrainian case described by Allina-Pisano (2008: 4), so in Khorezm entitlements to land 'exist in the world of bureaucracy and law'. In that world, to use the words of a district official regarding legal disputes in agriculture, 'law can be a weapon'. It is easy for authorities to find a legal reason to close down a farm and confiscate its goods or to change its terms of trade, thus pushing it into bankruptcy.

The Reshaped Command Hierarchy

With the implementation of the third step of agricultural reform in 2008–2009, the way which Uzbekistan was moving towards the market became clear. Neither family smallholdings under pudrat agreements nor large, kolkhoz-size farms, but instead a controlled number of brigade- or middle-size farms, was planned to occupy the formerly collectively owned land. In 2004, however, it was already clear that decollectivisation was being accomplished in Uzbekistan not by true privatisation of land but by 'de-shirkatisation' (disbanding of the shirkats) and 'fermerisation' (reorganisation of the old production units into newly established individual enterprises).

Decollectivisation did not disrupt the command hierarchy of production but merely reshaped it. Whereas in the past the agricultural system was based on direct control exerted by the kolkhozes (and later the shirkats), nowadays the state has moved away from direct involvement in agriculture while maintaining indirect control through regulations such as compulsory cropping schemes and a monopoly over input and output markets.

Nor did the end of the shirkats disrupt the command structure of agriculture; it only reorganised it to some degree. Even where the shirkats no longer exist, their former boundaries remain significant. Although almost all buildings, all orchards, and some machinery were sold at auction, most of the collectively owned tractors, housed in the MTPs, were not. In this way the skeleton of the shirkat – reorganised as a public company but controlled de facto by the district hokim, who appoints the manager (*rais*) of the MTP – survived privatisation. The same persons who previously managed the shirkats are now managers of the MTPs. The command and decision-making structure inherited from the kolkhoz, based on the kolkhoz rais and executive staff plus brigades, is still in place, although the nature of its task has changed slightly. MTP staff monitor and exert control over the implementation of fermer production plans, while the fermers are personally liable for performing their share of the plan. That the former shirkat is still a significant administrative unit is also demonstrated by the fact that district data on farmers' production quotas and harvest outputs are filed under the name of the local MTP station.

Another effect of the reforms on the hierarchy of production and power has been the recalibration of relations between hokim, rais, fermer, and dehqon. The most significant effect of the dismantling of the shirkats was that land allocation moved to the district level; it is no longer a purely local (kolkhoz-level) business. Whereas previously the kolkhoz or shirkat rais had a say in cropping decisions and was the key player in negotiations over land allocation, fermers now interact directly with district authorities (table 2). The founding of fermer enterprises, as well as changes in farm status and size, requires the district hokim's signature, and cropping plans and changes to them during the agricultural year are agreed upon with the chief of the district agricultural department. He is the second most important official in the district after the hokim, especially in rural districts such as Yangibozor.

Table 2. The evolution of the command hierarchy in agriculture in Khorezm.

Size (hectares)	Kolkhoz-Sovkhoz (Soviet Period)	Shirkat (from 1998)	Decollectivised Agriculture (from 2003–2004)	Consolidation of Farms (from 2008–2009)
20,000	Ispolkom chairperson appoints kolkhoz and sovkhoz rais	District hokim appoints shirkat rais	District hokim appoints MTP rais and fermers	District hokim issues decisions on farms
1,000–2,000	Kolkhoz rais Sovkhoz rais	Shirkat rais	Rais of MTP or farmers' union	Rais of MTP or farmers' union
50–150	Brigade head	Brigade head	Large fermer	Large fermer
10–50	—	Small fermer	Small fermer	—
1–10	Kolkhoz employees	Pudratchis and small fermers	Farm workers and small fermers	Farm workers

As the reforms intensified, the district authorities directly took over the sub-kolkhoz level of land management. As long as the shirkats existed, soon-to-be farmers applied to the rais for land and agreed with him on the size, location, and cropping specialisation of the farm. Now, farmers sign contracts for crops and fertilisers directly with the district supply facilities, so they are less subject to the MTP rais, especially if they own a tractor. The changes have substantially reduced the influence of the rais, whose material basis of power has been cut off from land and fertilisers and circumscribed to the facilities of the MTP.

A reserve of arable land (in 2004, some 5 to 10 per cent of total area, diminishing annually) remained allocated to the MTPs and was managed the way shirkat land had been managed previously. The overwhelming share of land, however, is now in the hands of fermers.

Decollectivisation and Local Governance

Along with decollectivisation, Uzbekistan has reformed the structure of its civil administration. For decades the administration of rural areas had been characterised by a double, civil and economic structure of management that was rooted in the double governance structure introduced with the cohabitation of the *sel'soviet*, for civil administration, and the kolkhoz, for economic administration, during the Soviet period. This structure lived on in postsocialist Uzbekistan in the form of an administrative division between the shirkat and the village administration, now called the *fuqarolarning yig'ini* instead of the sel'soviet.

The system changed in 2006. When general elections were held in June that year for heads of the eight village administrations in Yangibozor district, the former position of village administration chair (*sho'ro*) was replaced with that of village hokim (*qishloq hokimi*). This marked the end of the long-lasting administrative dualism, because it realigned local-level administration with the higher echelons of post-independence state administration. Village hokims originated in the same ideal of governance that was enacted at the district level with the district (*tuman*) hokim, at the regional level with the viloyat hokim, and on up to president of the entire nation. The model is one of state-society relations in which authority and state power are territorially defined, hierarchically conceived, and centrally organised (Massicard and Trevisani 2003: 216).

According to an employee of the Yangibozor district administration, or hokimiyat, the village hokim today unifies the functions of the village administration chair and the rais of the shirkat, thus simplifying and rationalising the exercise of power by the government at the former kolkhoz level of rural society. When I asked why this administrative reform was necessary, the interviewee expressed the opinion that after the end of the shirkats, fermers were not subject to the MTP chairmen as effectively as they were under the collectives in the past.[130] The impression I received was that because the power of the MTP rais over the fermers was essentially feeble, the local governance system was reformed as a means to restore his legitimacy and power by virtue of his affiliation with the state hierarchy.

This interpretation is validated by the fact that in six of the eight village administrations of Yangibozor, those elected as chairs were people who had worked as MTP raislar before the elections.[131] In Bo'zqal'a, it was the former sho'ro who became qishloq hokim, because, according to my source,

---

[130] Yangibozor, interview with R. J. Returning from a seminar (*pokaz*), he said that the hokim had complained that too few fermers were attending the meetings. Whereas in the kolkhoz period it was possible for all brigadiers and raislar to attend such meetings, this was hardly manageable after decollectivisation, with more than 700 fermers now dealing in cotton in the district. The hokim's conclusion, nevertheless, was that 'if so, then it was better with the kolkhoz. Everybody would have attended'.

[131] This happened in the Qalandardo'rman village council, which encompasses the Hamza and Bo'ston MTPs; in the Boshqirshayx village council (Jayhun and Xorazm MTPs); in the Oyoqdo'rman village council (Xalqobod MTP); in the Shirinqo'ng'irot village council (Shirinqo'ng'irot MTP); in the Bog'olon village council (Bog'olon MTP); and in the Chopolonchi village council (Madaniyat MTP).

the MTP rais at the time was attracted by more lucrative personal prospects and declined the district hokim's invitation to take the job. In the Chopolonchi village administration (with jurisdiction over Madaniyat MTP), where the MTP rais was appointed village hokim, the former sho'ro was made the new MTP rais, a switch that further illustrates how changes in positions and the denominations of public offices were nominal, while de facto pre-reform power relations remained unaffected.

Control and Coercion

During my fieldwork in 2003, it became clear after a few interviews that in Yangibozor the district was the level of decisive interface between local agricultural production and production parameters set by the government. It is therefore the district administration that should be seen as the crucial executor of government agropolicy. The agricultural system maintains the hierarchical organisation introduced during the years of the early Soviet Union, a hierarchy linked to the cultivation of cotton and wheat. Under the Soviet Union, the government supported local production with subsidised inputs and facilities, which were distributed and managed at the kolkhoz and district levels. Nowadays the trend is towards gradually abolishing such subsidies, although governmental and semi-governmental input supply structures and facilities remain.

For producers, however, these incentives do not make cotton sufficiently profitable. According to Ilkhamov (2000), the growing of cotton at the turn of the twenty-first century was still linked to a policy of coercion that reflected the central government's interest in ensuring that its cotton targets were fulfilled, and the district-level authorities were the most important executors of this policy. The coerciveness of the cotton policy as late as 2010 was partly hidden behind the state policy of economic programming, but it remained evident in the fact that fermers who produced something other than what was requested of them, or who repeatedly failed to reach their targets (thus accumulating debts and going bankrupt), lost their land and the money and other assets they had had to register in the name of their farm. This is a further difference between the kolkhozes and the newly established farms: collectively held enterprises could run debts, but today's fermer enterprises cannot have deficits for more than three consecutive years.

Every year, quotas for the so-called strategic state-ordered crops, wheat and cotton, are decided for every region (*viloyat*) in the country, and then planning and quotas move down to the district level. At that level the plan is again subdivided into new quotas and targets for the subordinate units – in the past, kolkhozes and shirkats; today, the MTPs and farmers' unions, and within them, the fermers. Above the district, at the viloyat level, the state authorities are more concerned with planning the reforms than with detailed monitoring of agricultural production. The details of the organisation of agriculture, as well as those of the implementation of reforms, are steered at the district level. The district hokim plays a key role in shaping the implementation of the reforms, and he both holds crucial responsibility for and is a key decision-maker in agriculture in general.

Plate 8. A seminar (*pokaz*) in which district authorities give instructions to *fermer*s on how and when to start sowing and fertilising.

At the district level, the agricultural cycle starts with the setting of production targets and the planning of production quotas at the beginning of the year. As the district hokim reiterated during the periodic seminars (*pokaz*) held for fermers, in 2004 the plan for Yangibozor was '27,000 tonnes of cotton and 5,400 tonnes of winter wheat: that is our contract. I have asked the [regional] hokim for more people, because we don't have enough in Yangibozor. During the harvest season, 4,200 persons will come for 30 to 35 days to help out'.

For every production unit in the district, the district-level authorities determine, for the current season, how many tonnes of output should be produced and how many hectares of land should be allocated to which crops. Just like the kolkhozes and shirkats before them, fermer enterprises are entangled in a state-steered system of input-providing and marketing organisations on which they depend for their production. This turns a fermer into a sort of small-scale shirkat, with similar links to the larger structure of agricultural production. As one fermer put it: 'There is no law that obliges you to grow cotton as a fermer. However, Uzbekistan is an agrarian state [*O'zbekiston agrar davlat*]. They establish the crop quotas for every viloyat, for every *rayon*, then for every MTP, and so on, for every fermer'.

In fact neither fermers nor district rural facilities are free in their decision-making. Fermers are not allowed to decide on contracts, and district input and output facilities have no say over their pricing and delivery system, which is regulated according to standards valid for the entire republic. Standards and rules exist for almost every aspect of agriculture, including the share of crops of cotton and wheat, the application rates of fertilisers, and the amount of water per hectare to be used for irrigation. However, norms can be superseded and standards disregarded, because the primary task is to meet the production target set in the plan. The district leadership has to take care that the district plan is fulfilled, and so the agricultural cycle still is an immense work coordinated from above. Cotton and wheat are matters of national interest (they are also called 'strategic crops'), and the hokim, supported by a chosen group of former shirkat-level, now MTP-level, managers and key district officials, in fact organises agricultural production from above, determining which percentage of land will be cropped in which way, which land will be exempted from the cultivation of state crops, who gets which land, who monitors production, and who carries out production.

Whereas in postsocialist Russia the former kolkhoz has increasingly been transformed into a 'hacienda' (Nikulin 2003: 144) – that is, the collective farm manager 'turns into a landlord', and the shareholders, into 'serfs' – in Uzbekistan, because of the cotton economy, this may already have been the case under the Soviet Union (Thurman 1999). Since decollectivisation, however, it is the district as a whole that increasingly resembles a (very) large farm, subdivided into MTPs and fermer enterprises, in which ultimately the district authorities charged with accomplishing the plan intervene at the lower levels of production. During the cotton-growing period in Yangibozor at the time of my fieldwork, it was not uncommon to see members of the 'cotton control group' (*paxta tekshirish guruhi*) monitoring the fields and giving instructions to fermers or, in their absence, to their farm labourers concerning the cultivation of cotton. Although the risks, capital investments, and decisions regarding farming belonged legally to the fermers, the farm labourers carried out the orders of the cotton control group, well aware of the relations between the fermers and the district leaders.

In Yangibozor in 2004, the district-level steering group amounted to some 20 key officials – heads of MTPs and of district-level agricultural offices – with whom the district hokim worked closely. During the intense periods of the agricultural season, they held meetings twice a day, morning and evening, to take decisions on the district's agriculture. After decollectivisation, district civil authorities were involved in agriculture and held responsibilities more than before. The monitoring of crops fell to this group, along with providing district fermers with sufficient inputs at the right times. In Yangibozor, local producers were merely executors of the production plan and essentially depended on the 'terms of trade' they agreed upon with the district authorities. Although a fermer was relatively autonomous in local farm management, farm liberties could be (and were) restricted at any time if the targets of the economic programme were endangered. These circumstances explained the need for strong supervision on the part of the district authorities, for whom strict control and monitoring of local producers throughout the period of agricultural production was necessary.

Clearly, decollectivisation of the agricultural sector was linked to attempts to contain the erosion of the state-mandated production plan and to enhance productivity by strengthening producers' accountability. Parallel to this process, the district authorities attempted to strengthen their control over agricultural production relative to the time of the kolkhozes and shirkats.

With decollectivisation, the re-parcelling of the agricultural land of the former shirkats augmented the district authorities' possibilities for controlling local-level land use and reduced the space for 'hidden land', or land that kolkhoz and shirkat authorities managed to keep off the official books and to administer outside the control of the state (for instance, by illegally selling its usufruct for a season to those who could pay for it). In the kolkhozes and shirkats, the units of production (the brigades) were very large, and accounting was done in an aggregated, shirkat-level form. Nowadays, fermers have to report their land use and production to the district statistical office, and they face legal harassment if they report wrongly. A special department of the district prosecutor's office deals with the monitoring of private farm activities. Forms and structures of land usufruct that previously were hidden now emerge through the constitution of fermer enterprises and become visible to official statistics through fermers' reporting.

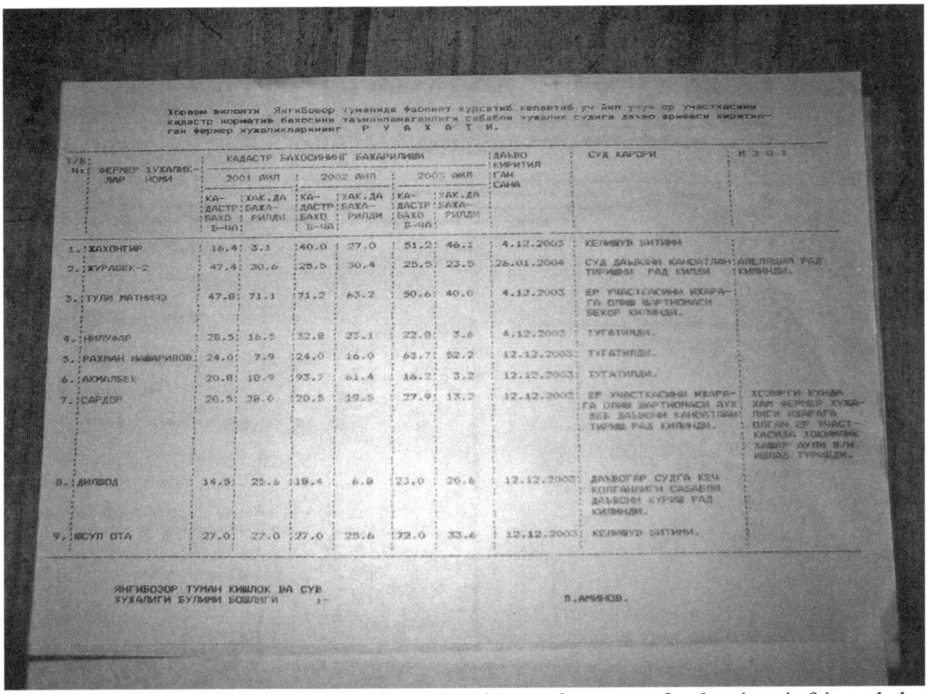

Хоразм вилояти Янгибозор туманида фаолият курсатиб келаётиб уч йил учун ер участкасини кадастр норматив бахосини таъминламаганлиги сабабли хужалик судига даъво аризаси киритилган фермер хужаликларининг Р У Й Х А Т И.

Т/Б №	Фермер хужаликлар номи	Кадастр бахосининг бажарилиши: 2001 йил, Кадастр бахо б-ча	Хак.да бажарилди	2002 йил, Кадастр бахо б-ча	Хак.да бажарилди	2003 йил, Кадастр бахо б-ча	Хак.да бажарилди	Даъво киритилган сана	Суд карори	Изох
1.	Жахонгир	16.4	3.1	40.0	27.0	51.2	46.1	4.12.2003	Келишув битими	
2.	Журабек-2	47.4	30.6	25.5	30.4	25.5	23.5	26.01.2004	Суд даъвони каноатлантиришни рад килди	Апелляция рад килинди.
3.	Тули Матнияз	47.8	71.1	71.2	63.2	50.6	40.0	4.12.2003	Ер участкасини ижарага олиш шартномаси бекор килинди.	
4.	Нилуфар	28.5	16.5	32.8	23.1	22.8	3.6	4.12.2003	Тугатилди.	
5.	Рахман Машарипов	24.0	7.9	24.0	16.0	63.7	52.2	12.12.2003	Тугатилди.	
6.	Акмалбек	20.8	10.9	93.7	61.4	16.2	3.2	12.12.2003	Тугатилди.	
7.	Сардор	20.5	28.0	20.5	19.5	27.9	13.2	12.12.2003	Ер участкасини ижарага олиш шартномаси бекор килиш тугрисидаги даъвони каноатлантиришни рад килинди.	Хозирги кунда хам фермер хужалиги ижарага олган ер участкасида хокимлик хабар ахли б/и ишлаб турибди.
8.	Дилшод	14.5	25.6	18.4	6.8	23.0	20.6	12.12.2003	Даъвогар судга кеч колганлиги сабабли даъвони куриш рад килинди.	
9.	Юсуп ота	27.0	27.0	27.0	25.6	72.0	33.6	12.12.2003	Келишув битими.	

Янгибозор туман кишлок ва сув хужалиги булими бошлиги :- П. Аминов.

Plate 9. A list of *fermer*s closed down by the *prokuratura* for having infringed the law.

Since decollectivisation, the district authorities' control over fermers' land use has been strict and, as far as I could observe, more effective than in the past. The tightening of control occurs on many levels. It is visible in the stricter supervision exercised through control agencies and militia and, more proactively, through the use of threats and the closure of farms to maintain an atmosphere of fear and competition among fermers. One interviewee told me, 'In the ministry the order is to "avoid fermers sowing too little cotton". Then it trickles down the ladder. Finally, once it arrives in the district: "Find me three fermers and shut them down. You have until 7 pm". That's how it works'.

Chapter 5
Decollectivisation, Labour Relations, and Kinship

One year after the disbandment of all *shirkat*s in Yangibozor, the employees of the local branch of the Fermer and Dehqon Association explained to me that, for the working population, the main difference between farming before and after decollectivisation was the switch from the family-based sharecropping agreement (*pudrat*) to the formal lease (*ijara* contract) governing land usufruct. For peasants (*dehqon*s), one effect of the dissolution of the shirkats was that they could no longer make pudrat arrangements. Dehqon households obtaining ijara farmland became fermer households. Those who did not become fermers had to work for fermers, and their working conditions changed. Families who did not become fermers were incorporated into new fermer enterprises, which, in order to avoid disruption of their cotton cycle, had to take over some of the shirkats' pre-existing arrangements.

The gradual introduction of a more individualised agriculture centred on fermers has had strong effects on the local organisation of agricultural labour. The reforms also had a watershed effect on the ways in which households related to agriculture, depending on whether household members had access to farm land as owners or in affiliation with owners or simply worked on farms as unrelated labourers. In this chapter I look inside the stratification created by decollectivisation by focussing on changed labour conditions and exploitation and by addressing what it takes to be a successful player in the context of decollectivisation – among other things, the mobilisation of family networks.

I also discuss how, contrary to the assumption that rural actors in Uzbekistan may be regarded as mere recipients of a top-down reform policy, strategic actions on both sides result in a local struggle around fermers' cropping patterns. Because of the peculiarities of Uzbek agricultural reform, fermers are not solely either market-oriented profit maximisers or subsistence-oriented producers. Because heavy constraints on farm liberties persist, unquestioned conformity to directives from above coexists with efforts to maximise profits regardless of those directives. Successful farming is a balancing act between the two extremes. The logic of farming follows the locally determined criteria of optimising integration into the still-preserved command system and trying to enhance the terms of usufruct relative to the command production apparatus.

New Cleavages in the Former Kolkhoz

Decollectivisation was an important moment of change for those involved in agriculture, because entitlements to land passed to individuals (affiliated with households), and land was redistributed to a minority of fermers. The majority of the former kolkhoz members were excluded and became simple dehqons, without direct access to the main part of the formerly collectively held land. In Yangibozor, fermers count as a privileged group, as is evident in the numerical discrepancy between fermers and dehqons. Fermers themselves, however, differ widely in their economic capabilities, status, and future prospects.[132]

As a result, decollectivisation has produced two different sorts of cleavages within the villages of the former kolkhozes. A first divide has opened up between those who obtained and those who could not obtain leases for farm land. A second has opened up among fermers themselves. While all fermers gained possibilities unavailable to dehqons, for some the new status entails greater risks and liabilities, and for others, fewer risks and greater opportunities.

To illustrate the scope of these new divides, I rely on two maps that I commissioned from Zarif, the owner of a large fermer enterprise in Xalqobod and, at the time of my interview with him, land measurer for the Xalqobod *motor traktor parki* (MTP), showing the patterns of land distribution in his village before and after decollectivisation (maps 5 and 6). Besides managing his own fermer enterprise, Zarif's daily occupation was to produce the papers for land leased to fermers and to measure and define land parcels for households requesting subsidiary plots (*tomorqa*) or land on which to build houses. Because of his work, he was well acquainted with Xalqobod's figures and reality.

---

[132] This last aspect became manifest most recently with the government's farm consolidation policy of 2008–2009, which significantly reduced the number of fermers from those originally established with decollectivisation.

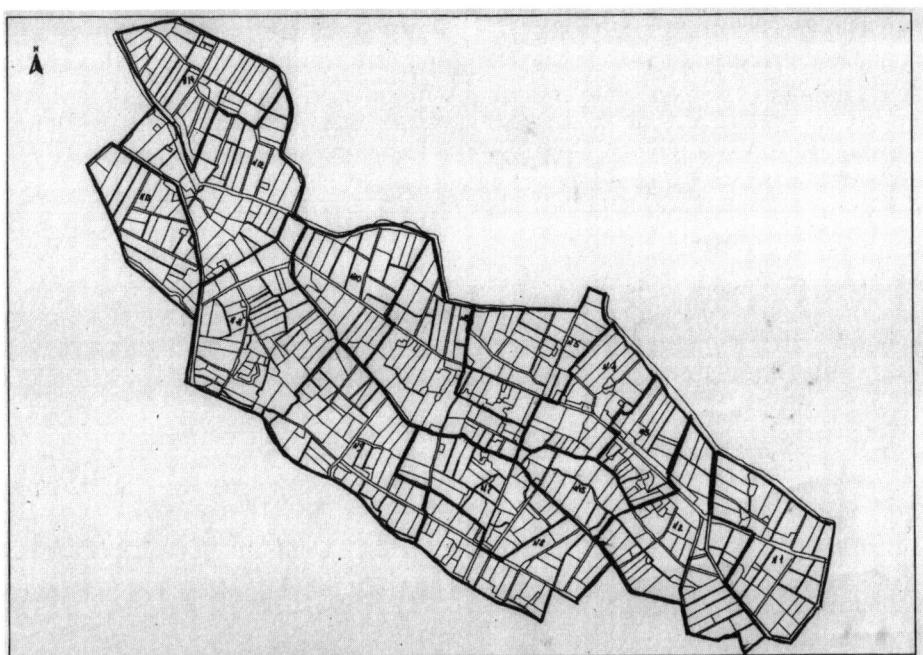

Map 5. *Kontor*s, or plots of land, worked by the brigades of the Xalqobod *shirkat* before decollectivisation. Adapted from ZEF/UNESCO Khorezm project.

Map 5 shows the boundaries of the farm land worked by each of the 15 brigades of the Xalqobod shirkat before decollectivisation. Each brigade was responsible for a varying number of plots (*kontor*s) – on average, 20 to 30 – each measuring within a range of about 3 to 15 hectares. At first the brigades worked the plots jointly, until the kontors were partitioned and worked by families under the pudrat system. According to Zarif, in the past as many as 1,000 people were employed in the shirkat's 15 brigades. In January 2003, most of the land formerly held by the brigades passed to fermer enterprises (map 6).

According to Zarif, his village, Oyoqdo'rman, which was co-extensive with the MTP and former shirkat Xalqobod, had approximately 10,000 inhabitants in 1,500 households at the time decollectivisation was implemented.[133] His figures show that 1,897 hectares of irrigated land then passed to 134 fermer enterprises (9 per cent of households). The MTP retained 106 hectares, which, like the fermers' land, were planted in state crops. Only 205 hectares of irrigated land were made available to the whole of the dehqon

---

[133] According to Raistat statistics for 2002, the Xalqobod shirkat (the Oyoqdo'rman village council) covered 2,681 hectares of land, 1,371 of which were irrigated and thus suitable for agriculture.

households as tomorqa smallplots. Thus, in Xalqobod a small minority of households received the overwhelming majority of the formerly collectively held land, and everyone else ended up with nothing but small tomorqa plots. There, decollectivisation manifestly created a deep cleavage between those who got land and those who did not.

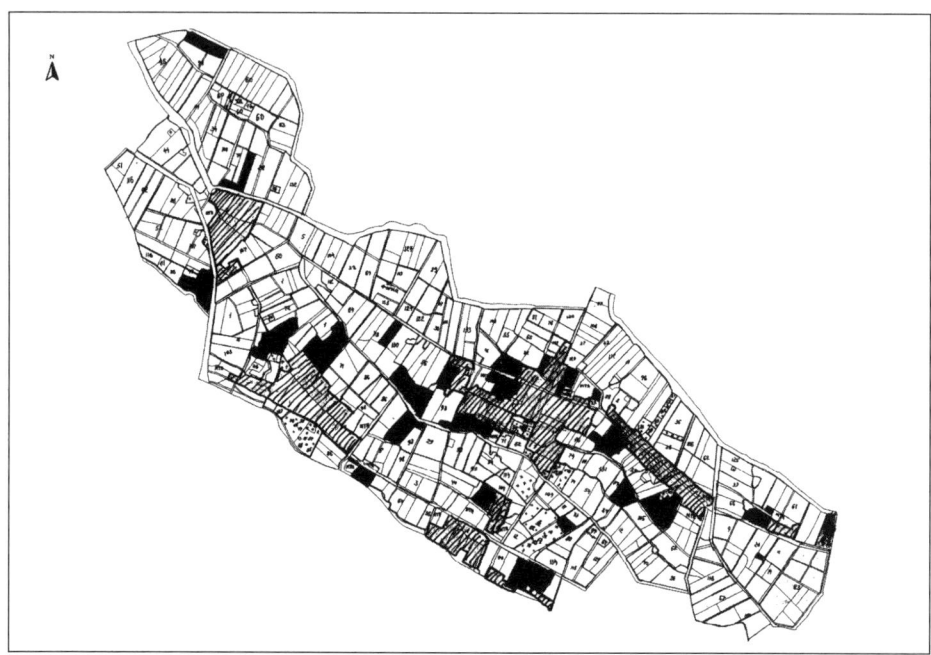

Map 6. *Fermer* enterprises in the Xalqobod MTP after decollectivisation, shown as the white areas with numbers. Areas in dotted pattern are the *shirkat*'s orchards, which were privatised as small *fermer* orchard enterprises. Areas in black are irrigated plots cultivated by *dehqon* households (*tomorqa* plots). Striped areas are those covered by settlements (*posolka*). Adapted from ZEF/UNESCO Khorezm project.

The maps for Xalqobod-Oyoqdo'rman show clearly how a situation in which the implicit status distinction between brigadier and *kolxozchi*, invisible on map 5, has changed to one in which the distinction between fermers and dehqons is sharp. Dehqons make up the majority of the village's population. They have less access to land and different usufruct conditions from those of fermers. It is fermers who have inherited the brigades' role in agricultural production, although in 2004 only a small minority of fermers had holdings as large as those of the former brigades.

Even those who received land in Xalqobod-Oyoqdo'rman represent a heterogeneous group. As seen in chapter 4, fermer enterprises vary significantly in size and type. In Xalqobod, decollectivisation followed the general pattern of the district: among the 134 farms, 42 (31 per cent of all farms in the MTP) were orchards with fewer than 5 hectares of land each. Among the other 92 farms, 51 (38 per cent) covered 5 to 20 hectares; 29 (21 per cent) ranged between 20 and 40 hectares; 11 (8 per cent) covered 40 to 80 hectares; and the largest farm, controlled by the former *rais,* amounted to almost 116 hectares of land.[134]

For comparison, the neighbouring Bo'ston MTP counted 393 households and 1,573 inhabitants in 2004.[135] With only 828 hectares of agricultural land, it was one-third the size of Xalqobod. There, fermer enterprises were granted to 52 families (13 per cent of households). Of those 52 farms, 40 (73 per cent) were registered as having small lease holdings of up to 20 hectares; 11 (22 per cent) ranged between 20 and 40 hectares[136], and 1 farm (2 per cent), with 54.8 hectares, was larger than 40 hectares. In both Xalqobod and Bo'ston, a handful of fermers held the largest leases to former shirkat land, and the majority held fewer than 20 hectares apiece.

Both divides, the one between fermers and dehqons and the one between small and large fermers, have been consciously established by the Uzbekistan legislature and district authorities. As a result, a new, almost class-like hierarchy is being created in rural society on the basis of differential entitlements to land. These characteristics were still incipient after decollectivisation, but although I could not conduct fieldwork after 2008, they seem likely to have become more pronounced after farm consolidation. Unlike neighbouring Kyrgyzstan, where the introduction of land ownership rights has prevailed in the course of post-Soviet agrarian reforms (cf. Anderson 1999), in rural Uzbekistan reform has privileged the maintenance of a state-driven, hierarchical principle regarding access to land.

With the impossibility of reviving the pre-kolkhoz land tenure regime in Uzbekistan after independence, the end of collective agriculture there led to redistribution instead of restitution of formerly collectively owned land and other assets (cf. Verdery 2000). Although land distribution does not emerge from nowhere, but is carried out within a framework of previously created parameters, still it creates a moment of discontinuity with the structures of the past. In Uzbekistan, as becomes clear from Zarif's maps, decollectivisation did not simply follow the boundaries of the past. With fermers' and dehqons' differing entitlements to land, the work of the land commissions introduced into the rural context a new stratum of boundaries, parti-

---

[134] Source: Yangibozor harvest balance sheet, 2003.
[135] Source: Qalandardo'rman village council population count, 1997–2004.
[136] Source: *Yangibozor tuman 'Bo'ston-Yangibozor' mukobil MTP va uning hududida joylashgan fermer xo'jaliklarning 2004 yil ekish rejasi* [Crop growing plan of the fermers affiliated with the MTP Bo'ston], 2004.

tions, and segmentations, which sometimes overlapped with and sometimes crossed the older stratum of boundaries of the Soviet kolkhozes, brigades, and kontors.

Because of the sensitivity of the topic and the transitoriness of the studied context, there were many questions I could not ask, and for other questions it was impossible to find answers. For example, it remained unclear to me to whether the planners of decollectivisation had purposely shaped land entitlements to create a pyramidal hierarchy of first-class and second-class fermers, or whether that outcome was instead the result of a discrepancy between the reform demands of authorities at the highest state levels and the limits and contingencies of the local level. Such discrepancies had often been at play in Yangibozor even in the late socialist kolkhoz (chapter 3). Some officials I interviewed mentioned that fermers' lots were originally meant to be bigger but in the end had to be reduced because of the shortage of capable applicants. This suggests that the outcome should be understood as a compromise between local capabilities and state desiderata. On the other hand, the turn of events of 2008–2009 seems to suggest some sort of premeditation of the outcome of decollectivisation, which finds corroboration in the fact that all MTPs for which I could obtain information in Khorezm in 2004 followed roughly the same distributional pattern.

The new social cleavages opened up with decollectivisation relate to the unequal conditions of dehqons and fermers in regard to access to land, labour, and, more generally, the possibility of gaining one's livelihood through agriculture. Before exploring these aspects in subsequent sections, I must mention the importance of kinship relations, which pervade labour relations, farm management, and land 'ownership' or access to land among fermers and dehqons alike. The divide between families who could access the resources of a fermer enterprise by virtue of their kinship relation with the owner of the farm, and who thereby joined in the risks and opportunities of the enterprise, and those who were excluded from land because of their lack of kinship relations with fermers obviously mattered a great deal in the decollectivised villages.

In conversation, a chairman (*sho'ro*) of a village administration estimated that, through kinship relations, perhaps 40 per cent of the families of his village in Yangibozor district had direct or indirect access to fermers' land, even though far fewer households owned fermer enterprises. My 2004 survey in the Bo'ston MTP suggested a lower figure. Out of a sample of 40 households selected randomly from the 393 households listed in the village administration registry books, only two owned fermer enterprises, and only two others had close kinship ties with another household that owned a fermer enterprise. In another conversation, rural residents in Yangibozor stressed that despite hardship, 'no one in the village starves', but many families came close to it and lived by 'tea and bread alone', getting to 'eat meat only at *to'y*s [celebrations]' (cf. Zanca 2010: 100). In this respect, the divide created between kinship networks in the villages by differences in access to land was perhaps even more significant for rural people's livelihoods than the divides between fermers and dehqons and between large and small fermers created by the implementation of decollectivisation.

The Worsening of *Dehqon*s' Livelihoods

Commenting on the end of the kolkhoz, a dehqon explained to me the difference between working for the kolkhoz and working for a fermer:

> Now you work more. In the past the *kolxozchi* were organised in brigades, 15 to 18 for each kolkhoz, 30 to 40 persons in each brigade. Often, the neighbours of a street, an *elat,* an *avlod,* used to work together on some 100 hectares or so of cotton. Work time was regulated. You would start at six, you would take tea [*choy*] after two or three hours of work, work again, and then you would have lunch. After lunch, there would be some more work and then back home. Today the *rabotchi*s [workers] have to agree with the fermers.[137] They walk from the village to the field early in the morning, and walk back at late evening.

In this dehqon's account, less regulation of labour concealed a worsening of labour conditions. A dehqon's belongings and the land he worked on were the determining factors in his economic situation, and this was the main difference with his employment in the Soviet kolkhoz, for which he received a reliable income in the form of a salary. Generally, the dehqon was now much more dependent on his agricultural produce and more exposed and vulnerable to health problems, water shortages, and harvest failures than kolkhoz workers ever were.

Heating, too, became a major problem, because during the very cold winters gas was no longer delivered, and there was nowhere to get firewood for free. Since the 1960s, when the mechanisation of agriculture and irrigation triggered the expansion of agricultural land, peasants' access to firewood from the forests and fallow land around Yangibozor had been progressively reduced. One reason for working for cotton fermers was that after the harvest dehqons could gather the dry cotton plants and use them for cooking and heating throughout the year.

---

[137] *Rabotchi,* a Russian word with an Uzbek suffix, means 'worker' and is used synonymously with the Uzbek term *ishchi.*

Another detrimental change for dehqons was that the previously lucrative practice of grazing privately owned cattle on kolkhoz and then shirkat land became impossible after decollectivisation. In the past, cattle were especially valued for dairy products and as form of savings to be cashed in when households faced major life-cycle expenditures such as those for dowries, wedding celebrations, and new house construction (Rasanayagam 2002b: 67, 193–194). City dwellers envied villagers this economic opportunity. Since decollectivisation, fermers have claimed exclusive grazing rights on their newly obtained land and guard it jealously, so dehqons have nowhere to graze their cattle – a source of tension between the two groups. The number of cows per household in the villages of Yangibozor has progressively diminished since independence (cf. Aw-Hassan et al. 2004), from 10 to 15 during the late Soviet period to only a few in most cases today. Only fermers with their own land can obtain enough fodder to keep livestock in any quantity.

Garden products, too, were important components of rural families' incomes during the kolkhoz period, but unlike cattle, they have now become essential for ensuring households' livelihoods. A dehqon working on his tomorqa in Yangibozor city spoke spontaneously about this situation:

> I am worker in the cotton ginnery [*paxtazavod*]. But the only thing that brings in money is this small plot of land [showing the land he was working on] and 7,000 *so'm* per month from the cotton ginnery during the cotton harvest. Even so, they don't pay cash. They give us a piece of meat instead. All I get to eat comes from this plot. This is the second seeding I'm doing this year: vegetables. I harvested the winter wheat a month ago. There is nothing here. The only factory we have here is the cotton ginnery. Forty per cent of the population works on construction sites away from here or goes to Russia. They even go and come back for only $100.

In the Khorezm region, irrigated plots yield two harvests: winter wheat, harvested in June, followed by a fast-growing variety of rice that is harvested in September or by other crops if irrigation is insufficient for rice. According to calculations I made together with staff members of a shirkat, the average income for a well-performing tomorqa in 2004 was about 120,000–150,000 so'm (at that time, 1,000 so'm equalled US$1) for the first harvest and about 240,000–300,000 so'm for the second harvest. Subsidiary household plots usually had greater productivity than the large plots sown in state crops.

Plate 10. A *dehqon* with the wheat harvest from his *tomorqa* plot in front of his house.

Still, dehqon farms were too small and economically insecure to reliably feed a large household. Furthermore, although rural households could profit from the produce of their tomorqa plots, that land still could not be considered privately owned, as emerged during this conversation with a man named Xudoshukur in a village of Bog'ot *rayon:*

TT: Can't the people grow what they want on their tomorqa? Why is everybody growing rice?

H: They grow rice because they need cash, or otherwise they'll eat it. Also, if your neighbour grows rice, you can't grow anything else, because of the rising groundwater level. In order to grow a crop you have to agree with your neighbours.

TT: How do people decide?

H: The rais decides what the crop will be. Usually, if possible, it will be rice. The [plot allocated for use as a] tomorqa will change with rotation after some years, but the size will remain the same: one tomorqa for 10 to 15 people. In a year this can give you up to 800 kilograms of wheat and 800 kilograms of rice, if you have done your work properly. Additionally, because the families work as *pudratchi* on a field, they'll get 120,000–130,000 so'm a year out of that. But not cash. Instead, they'll give flour, wheat, oil, such things. They say in the offices there is no cash. The problem is that if you need a doctor or you need to buy medicines, or if you make a *to'y*, you will need cash. People are having troubles.

In other words, although people are not under obligation to grow state-determined crops on their tomorqa, they do experience external constraints on what they should grow, whether from neighbours or from the rais. Because of this, and especially because of the small size of tomorqa plots, that land forms an insufficient basis for regarding dehqon farms as economically viable and for considering the dehqon a sort of small fermer (as a district official suggested in a speech made during the time of my fieldwork). A tomorqa is too small to feed an average dehqon family. For dehqons, the safety net once provided by the kolkhoz has vanished. If a breadwinner becomes ill, a rural family's livelihood can be threatened in a way that was previously unknown. A dehqon family without its own assets (house, orchard, cattle) is more vulnerable than it was during the kolkhoz period.

Changing Labour Relations

Relationships between fermers and their workers, I observed, were ambiguous and differed from farm to farm. Depending on the fermer's personality and farming arrangements, labour relations could be more or less egalitarian. Fermers felt varying degrees of paternalistic responsibility towards their workers, who in turn, depending on their relationship with the fermer, were involved to varying degrees in the fermer enterprise. Most fermers lived in conditions not too different from those of many dehqons, while dehqons often maintained linkages with fermers based on kinship or on the work habits of the collective. In Bo'ston I attended the wedding of the daughter of a former rais to an army sergeant who was the son of a simple tractor driver. My table companions stressed that there was nothing special about this, rectifying my assumption that a rais would not marry his daughter to someone who did not belong at least to the group known as 'large farmers' (*katta fermer*s). In the aftermath of decollectivisation, the relationship between fermers and dehqons resembled a class relationship in the making. Most fermers, especially if they owned small farms, were not yet fully amenable to considering themselves a separate class of agricultural producers, but relations were moving quickly towards more definite forms, rapidly eroding the still indefinite middle ground between reform 'winners' and 'losers'. Partly because farms were newly established, and partly because farm sizes and specialisations had not yet stabilised, the situation was in flux.

In the following sections I illustrate relations between fermers and their hired labourers, relying mainly on examples from a cotton farm in Madaniyat and a rice farm in Jayhun.

Labour Relations on a Cotton Farm

Mo'hammad was the supervisor (*ish yurutuvchi*) of the cotton farm in the Madaniyat MTP on which, together with my colleague Kirsten Kienzler, I carried out interviews while conducting a field experiment with fertilisers during spring and summer 2004. Known as 'Rahimboy dev', the farm was owned by Mo'hammad's elder brother, Hayrullo, a former agronomist for the kolkhoz; the two were members of a 'territorial-kinship group' (Snesarev 1976) nicknamed Dev. During the wheat harvest that year, I asked Mo'hammad how his workload had changed since he had become the farm manager. He gave me the following answer: 'Unlike during the shirkat, I am always running, from one corner of my farm to the other. Every day I walk 30 kilometres. It reminds me of my military service in Leningrad. I have to take care of everything on the farm. For instance, now the wheat harvesting machine [*kombayn*] of the MTP is broken. If it doesn't get fixed in time, I'll lose my turn. How will I harvest then? But if I don't fix it, no one will at the MTP, and so comes extra work for me'.

During that cotton season I gained some familiarity with the farming arrangements on Hayrullo and Mo'hammad's enterprise. In 2004 the farm encompassed 26.9 hectares, 5 of which were planted in wheat and the rest in cotton.[138] Mo'hammad supervised the work of 3 farm labourers (*yo'llanma ishchi*s) and 16 *gektarchi*s.[139] The difference between the two, explained Zarif, Hayrullo and Mo'hammad's elderly uncle and one of the gektarchis on the farm, was that a gektarchi tended one hectare of cotton and had no other duties. At the end of the harvest he received in payment the cotton stalks from the fermer and the cotton oil from the state. A yo'llanma ishchi, in contrast, worked for the fermer, 'all over the farm. What the fermer orders him to do, he will do [*fermer ne ish buyursa shuni qilatti*], for instance, water, tractor, anything. They agree on the payment among themselves'.

---

[138] The cadastre map showed 28 hectares, including the fields, streets, and so forth.

[139] Dehqons in Yangibozor used the term *gektarchi* as a synonym for *pudratchi,* although strictly speaking the two terms meant different things. During the shirkat period, *pudratchi* referred to land rental arrangements for the growing of state crops, under which a pudratchi received several hectares of land to look after. After the shirkats were dissolved, the term became more applicable to the role of fermers. A gektarchi works for a fermer and does not have same legal status as the pudrat assignee in the shirkat. He is not involved in the handling of agricultural inputs.

Ravshan, one of the three yo'llanma ishchis, explained to me what the work of a gektarchi consisted of: 'You have to weed the field several times, you have to irrigate the furrows, pick away the tops of the cotton plant, so that it does not grow too much in height, because otherwise there will be fewer cotton bolls. All the work that is not done by tractor must be done by hand'. It is hard work, from spring until harvest, which begins towards the end of August. In the first year of farming, Mo'hammad's and Hayrullo's families decided to keep all the land for themselves and to divide the work among their extended family. Mo'hammad recounted: 'I took five hectares, my brother took five, and so on. But it was too much to be looked after. For instance, you need to do the weeding within two days. If you don't have enough workers, you can't make it. We decided to take workers'.

The Rahimboy dev farm consisted of five plots, called *otiz* in the local dialect (*kontor*s in the language of the land measurers). Each had its own name and had been divided among the 16 gektarchis and their families. Two plots, of six hectares each, were called, respectively, 'Qosim taqir kontor 1' and 'Qosim taqir kontor 2', after a former mullah and landowner who had lived in the area before the war. He was a bey unrelated to the Dev, as Zarif, the old uncle, explained. Zarif, Mo'hammad, and 20-year-old Ravshan, the yo'llanma ishchi, shared the six hectares of a plot called 'Yog'ochqa', which was the plot on which the kolkhoz had grown fodder crops in the past. The plots 'Uchmuyush' (meaning 'triangle', because of its triangular shape) and 'Sho'r otiz' (salty plot) were divided among other gektarchis, some of whom were related to the Dev and some of whom were not. Among the unrelated gektarchis, some were neighbours from Hayrullo's street, and others came from a neighbouring village.

Zarif highlighted the difference: 'The relatives work to help the farm. The unrelated work for the cotton'. When I asked for an explanation, he said, 'Take Ravshan, for instance: he came and asked for work. They didn't know each other before. He asked Hayrullo, "Give me cotton [*paxta ber*]". Among us, when one says this, it means that one wants to work on the field, meaning that he will work until the cotton picking, and after that he'll take the cotton stalks for firewood. With us it is different. We are relatives, we work for the farm'. Ravshan's father, a former tractor driver for the kolkhoz, fell severely ill in his mid-forties and had to stay home, taking early retirement. Out of economic need, Ravshan, still unmarried, took over responsibility for the family and decided to seek work at Mo'hammad's fermer enterprise, which he had heard about in the village. Unlike many of his peers, Ravshan remained in the former kolkhoz instead of seeking work abroad, in order to remain close to his parents.

The cotton stalks (*poya*) left over after the harvest, which dehqons needed to fuel cooking fires, were an unofficial but greatly needed remuneration. All gektarchis, both related and unrelated, took them from their part of the land after the harvest. Mo'hammad considered the granting of cotton stalks not as a payment but as a form of acknowledgement for the gektarchis' hard work: 'They have worked, they have grown the cotton. As a sign of respect [*hurmat uchun*] we let them take the *poya* from the fields of our farm'.

Payment arrangements on the farm were not transparent, and discomfort and even shame over talking about the issue were evident. 'Cotton is good work' was a recurring answer from one of the gektarchis, but it obviously was untrue. Besides official payments, stipulated on paper and translated into some non-perishable foodstuff from the state retail shops, said Ravshan, each gektarchi had an individual agreement with Hayrullo: 'Some ask for money, some ask for land, some ask for a share of the harvest'. Cash arrangements, unusual for cotton in any case, were not made on Hayrullo's farm. Sharecropping was more common. Ravshan, for instance, asked for some land to plant after the winter wheat harvest, 'if there were any seeds left, and if the *raislar* let Hayrullo plant the land'. Ravshan would then divide this second harvest with the fermer. The Devs planted a small part of the gleaned wheat plot for themselves, in 10 *sotik* (a tenth of a hectare) of vegetables and 30 *sotik* (3,000 square metres) of rice. Many fermers saw the possibility of a second harvest following winter wheat as a way to remunerate their labourers through sharecropping arrangements.

During the kolkhoz period, farm land was allowed to rest from time to time. In 2004 some fermers told me that land was now being excessively exploited. Hayrullo explained that the district authorities might impose fallowing out of concern for the over-use of soils or because of water shortages, thus possibly undermining arrangements made between fermers and their workers. Fermers whom I interviewed repeatedly said, 'Land has become like a drug-addict [*Yer narkoman bo'p ketdi*]', meaning that it needed more and more fertiliser but still produced less. The sentence was eloquent not only about the over-exploitation of soils but also about the thoughtless use of inputs and the mounting pressure to attain large harvests. As Xusaynboy, another fermer from the Madaniyat MTP, put it:

>During the kolkhoz the minimum wage was, say, 50? Well, we received 100. We could make a comfortable living. Then, during independence, wages fell, people started not to bother any more about the land, so the collectives did not yield profits anymore. Now the farmers are taking care of the land again, things will improve. Things are better already. But still, if in 1985 we reached 35,000

tonnes of cotton [in our district], this year the target is 28,000. We still haven't yet reached the level of the past.

As during the time of the brigades, farm labourers still join in the harvest of the state crops, cotton and wheat. On Hayrullo's farm, the winter wheat harvest in 2004 amounted to 20 tonnes, 9 of which were delivered for plan fulfilment, and 8.9 of which were distributed among the farm labourers, both related and unrelated, gektarchi and yo'llanma ishchi alike. The rest was, in theory, at the fermer's disposal. In reality, two tonnes of wheat had to be delivered for donation (*xayriya*) for the renovation of the district's hospital, as part of the fermers' duty to the community – something I discuss in greater detail later.

The Cotton Harvest

After the collectives became unable to pay salaries, cotton picking became the only opportunity for farm workers to earn cash. The year 2004 brought an excellent harvest, and many cotton pickers (*paxta teruvchi*s) were needed on the Rahimboy dev farm. During the harvest period Mo'hammad spent his days weighing the cotton collected by the labourers in the fields and his nights delivering it to the district cotton ginnery, where all fermers of the district queued up to do the same. At night the ginnery became a stage where all the fermers of the district met and compared their performances, in a competition for the right to invite colleagues to celebrate the fulfilment of the cotton target. On Mo'hammad's farm, the cotton picking team consisted of approximately 40 people, who together picked the cotton on a single plot and then moved to the next. It took three to four days for the team to accomplish the first round of picking, after which they started again from the first plot, for in the meantime the cotton bolls that were still closed during the first picking had opened.

According to an official of the Fermer and Dehqon Association from Madaniyat, during the 1980s, 60 to 65 per cent of all cotton was harvested by machines. Nowadays, he added, 'there are no machines. The fermers don't use them, because they are expensive. You have to pay for the tractor and the fuel. It's good to pick cotton by hand. It's cheap, and the quality will be better'. On Hayrullo's farm the cotton pickers were mostly young boys and girls with their mothers, along with some men, such as Ravshan. Each cotton delivery was weighed and recorded separately. Payment in September 2004 was 30 so'm per kilogram. I was told that a good cotton picker, able to harvest 100 kilograms a day, could earn 3,000 so'm (US$3) a day during the early days of the harvest. Later on, 80 kilograms a day was a good result, which many pickers on Mo'hammad and Hayrullo's farm were able to reach only because some families pooled their pickings and counted them as one.

Gulorom, for instance, one of the gektarchis on the farm, together with her daughter, combined pickings of 10, 12, 19, and 16 kilograms of cotton on one late September day.

Plate 11. *Fermer* Mo'hammad weighing cotton in his fields during the harvest.

In the early weeks of the harvest the pay was enough to attract sufficient workers from the villages. Later, what was still called by the Soviet name *totalnaya mobilisatsiya* began, when all public offices and schools closed to allow the district's workforce to join the effort to meet the district's cotton production target. In Bo'ston and Madaniyat, as in the other village administrations of Yangibozor, the administration chairman, together with the *posbon* (a sort of village bobby who fulfils the tasks of a unarmed policeman and assists the militia in the villages and in the *mahalla*s or *elat*s), compiled lists using the village registry books to check how many family members could be counted for the mobilisation and how many children were too young to be recruited for cotton picking. Unlike dehqons, students and employees from state organisations who were recruited picked no more than 20 to 30 kilograms a day. By resolution of the district authorities, defoliants were routinely used on the fields during this later stage of the picking, in order to speed up the harvest. Drying out the cotton plants, defoliants blocked the development of the cotton bolls, with the consequence that from then on all the cotton could be harvested at once.

Because Yangibozor had fewer inhabitants than other districts of Khorezm, university students had been compelled since Soviet times to come help harvest cotton there when pay and daily harvests became insufficiently attractive to the kolkhoz workers. They were hosted in school buildings and assigned to various MTPs and then to various fermers. College students in the district and schoolchildren not younger than the ninth school year were also recruited. For college students, the cotton harvest lasted approximately 40 days, beginning in early October. Their attitudes towards compulsory cotton picking were mixed. Many tried to avoid going; others accepted it as an unquestioned necessity. Child labour, on the contrary, was a sensitive issue, about which both ordinary people and staff of the command hierarchy avoided talking.

People's statements about compulsory labour in the cotton harvest were ambivalent. One fermer, on whose land students from the natural science faculty of the University of Urganch worked for a week, justified the recruitment of the students on the grounds that he himself had picked cotton for the kolkhoz when he was a schoolboy: 'Among us, among our people, it is so [*Bizda, xalqimizda shunday*]'. Another fermer said, 'We don't need them [the students]; we can pick the cotton by ourselves'.[140] Yet he profited from a deal with a school director who assigned several classes to his fields early in the harvest, helping him reach his target earlier than other fermers.

A fermer stressed that growing cotton was convenient and offered opportunities that no dehqon could ever have in agriculture:

> If you are a fermer and you grow 20 hectares of cotton, at the end of the year, if you subtract expenditures for credit, labourers, inputs, you'll get maybe 5.5 million so'm. It's good money that no dehqon can see. Look at me. With 16 hectares of cotton I have to deliver 50 tonnes. I'll get maybe 11 million from the ginnery. From the state I received credit of 3 million at 5 per cent interest rate. Eight million are left. I have to pay the cotton pickers 1.5 million and pay in advance the expenditures for the winter wheat season. Next year I want to get less credit, and in a couple of years you will see that not only I but most fermers will walk on their own legs, without any help.

---

[140] I could not determine precisely how much of the district's cotton harvest had been collected by schoolchildren and university students. A fermer told me that a third of his cotton had been picked by the pupils of the nearby college. An *elatkom* official told me that about 40 per cent of the cotton in his MTP had been collected by either schoolchildren or college students.

This fermer's optimism was dashed by the policy of 2008–2009, but in 2004 a sense of joining a superior cause for the national interest by growing cotton still led many fermers to share with district authorities an interest in fulfilling the cotton plan. They expressed hope and enthusiasm about the newly gained prospect of managing their own enterprises.

For fermers and district authorities alike, speed was a decisive factor in the cotton harvest. For one thing, they feared that bad weather in the approaching winter would affect the quality of the cotton. For the authorities, fear of cotton theft was even more acute, so they urged fermers to deliver their cotton daily and not leave it in the fields. During the harvest, the militia patrolled the streets of the district day and night. A land measurer told me that it was a widespread practice for cotton pickers to take the cotton from a field and sell it to a fermer who offered better pay or who paid immediately. The reason for this, I was told by a colleague from our project, was that many fermers kept part of the money earmarked for the cotton pickers for themselves, because banks hampered the withdrawal of cash from fermers' bank accounts throughout the year, except during the cotton harvest period. Moreover, if a fermer had already met his cotton target, he might try to sell cotton undercover to other fermers who were in difficulty, although any such trade in cotton was prohibited by law. Some fermers claimed that the weighing at the cotton ginnery was rigged and that *yo'l puli* (literally, 'money for the way') was being extorted from them upon delivery. Stories and gossip about cotton theft, a sort of genre in their own right, circulated widely, but it was difficult to determine their validity.

All in all, fermers were ambivalent towards the issue of cotton picking. Many were aware that the rate per kilogram set by the government for the cotton pickers was too low and that the prices paid to fermers at the cotton ginnery and to pickers in the field were too distant from each other. A land measurer-cum-fermer summarised the point:

> At the beginning of the harvest you can pick up more, you make more money. In the follow-up picks you get less cotton. Therefore the defoliant is a blessing. You save time and money. If a dehqon has other things to do – for instance, if he has to prepare his tomorqa for wheat or he must beg a land measurer for a tomorqa or must harvest his own rice – he will not come to the picking. There will be a labour shortage. So you see, if rural people become too rich, no one will pick up the cotton.

Plate 12. Boys from a school class resting after picking cotton for a *fermer*.

Labour Relations on a Rice Farm

Together with my colleague Christopher Conrad, who needed large areas planted in rice for a remote sensing experiment, and thanks to the mediation of an official of the *boshqarma,* I got acquainted with Muxtor and his brother Xudoybergan, whose neighbouring rice farms encompassed 25 and 35 hectares, respectively. In 2004 I visited their farms frequently during the rice season, late spring and summer. Situated in the Jayhun MTP, not far from the Amudaryo, their land was unsuitable for cotton growing, so they specialised in rice. Rice had been grown in the area since the establishment of the sovkhoz in 1976, but its importance had diminished during the 1990s because of the crop policy prioritising cotton and winter wheat.

In 2004 rice prices were higher than during sovkhoz times, and the suspension of the state-mandated procurement quota on rice a few years earlier had helped increase the crop's attractiveness for producers. In Khorezm, rice farming requires far less work than cotton farming, although irrigation, weeding, fertilising, and harvesting at times require intensive labour. The rice season is relatively short, so rice can be sown after the winter wheat harvest. Muxtor and Xudoybergan did not raise winter wheat, however, and harvested one rice crop a year.

Muxtor employed four yo'llanma ishchis, two of them aged 19 and 20, respectively, and the other two younger. They had been recruited from Muxtor's village of origin in Urganch *rayon* and from the village nearby. The workers' pay was much more attractive than that of a cotton farm. 'If you work on cotton, they'll just pay you with foodstuffs, no cash', one of them told me. Each of the four received 15,000 so'm for every month of work during the rice season, up until September, and each was granted 200 to 300 kilograms of rice from the harvest. To weed the rice fields, Muxtor on one occasion had to employ 35 workers (*ishchis*) from the nearby village. He paid them 4,000 so'm per hectare and gave them their meals during their workdays. He hired extra workers again for the harvest and paid them in cash. Muxtor's farm had eight members (*a'zo*), including himself, and all the others were his relatives from the home village in Urganch *rayon*. Although they all received shares of the farm's profits, only one of the other members was directly involved in the fields. Roman, the husband of Muxtor's sister, had formerly worked as a veterinarian in a kolkhoz and now worked as Muxtor's farm supervisor. That meant he also looked after Muxtor's cattle, which grazed on the farm land.

Roman repeatedly emphasised his brother-in-law Muxtor's generosity and good reputation. For the wedding of Roman's two sons, Muxtor, without being asked, gave him 600,000 so'm as a present and brought all the vodka (*aroq*) for the *to'y*. Roman was remunerated by Muxtor through what he called a pudrat. Under this informal sharecropping agreement he received six hectares of Muxtor's land. Muxtor procured the fertiliser and made sure sufficient water was available throughout the season, and Roman delivered five tonnes of rice per hectare at the end of the harvest, keeping the rest for himself. His share of the crop could amount to 10 to 15 tonnes of raw rice, ready to be sold on the market. This was a remarkable amount of capital, which the labourers on the rice farm, who performed all the seeding, weeding, and harvesting, would never have been able to earn.

Relations between Fermers and Farm Labourers

On Hayrullo's cotton and winter wheat farm, relations between the farm-owning Devs and their unrelated workers were good, characterised by reciprocal *hurmat,* here meaning mutual respect as well as a shared understanding of the rules and roles on the farm. During the cotton harvest, jokes and a relaxed atmosphere eased the burdensome work. Related and unrelated people worked together, although they had different motivations for working. Nevertheless, Hayrullo and his brother were careful in monitoring the labourers' work: 'if no one looks at them, they will not work and go away', Hayrullo explained to me once during the weeding period. The farm labourers, on the other hand, knowing of my good relationship with the fermers, never expressed any criticism of their employers.

Mistrust between farm workers and farm owners was more evident on another farm. This was a large (60-hectare) cotton farm in Jayhun, owned by a 50-year-old former official of a state organisation (*tashkilot*) that had been closed down. Upon my visit, he asked me to keep quiet about the fertilisers I saw in the storage room on his farm, because his workers would steal them if they knew about them. None of his relatives worked on the farm; he trusted only his farm supervisor. When I asked why he did not involve family members in his fermer enterprise, he explained the advantages of hiring unrelated labourers: 'It's not so good to have too many family members among the workers of a farm. They work less than others. Among the yo'llanma ishchis, one gets the salary, and [his] family members, when there is a need, come along and work with him. So if you have 10 to 15 workers for 30 to 40 hectares of state plan land, when there is need to do the weeding, for instance, they will bring along their families to finish work earlier'.

On larger farms, especially those that were more commercially oriented, relations between fermers and workers seemed to be more anonymous and hierarchical. In a conversation I had with a group of farm workers whom I once asked for directions in Jayhun, I discovered that although they knew the size of the farm on which they worked and the name of the farm supervisor, they knew neither the name of the farm owner nor the official name of the farm. Later I discovered that this was not a unique case, although my evidence is fragmentary and rests on fermers' accounts rather than on direct observation. The impression I gained was that after decollectivisation, labour relations were less often than in the past embedded in relationships such as those of neighbourhood, family, and friendship. They had become increasingly impersonal, triggered solely by the need for work and a salary.

In this respect, Hayrullo's cotton farm, resembling a micro-kolkhoz, differed from Muxtor's rice farm, on which, thanks to its specialisation in a freely marketable crop and its better availability of cash, labour relations had evolved more towards the market principle. I also heard, however, of cotton farms that had sufficient capital to develop away from the 'gektarchi' type of labour and towards temporary hired labour, like that on Muxtor's farm.

The state has maintained a strong influence over the definition of labour relations on farms, although its influence is more indirect than in the past. The state 'retreats' to the setting of prices and parameters while leaving to the fermer the burden of being the land's master. The examples I have discussed show how the system is based on the exploitation of the non-kin workforce, an exploitation that has worsened since the dissolution of the shirkats. Agricultural labourers hired on a daily basis – *mardikor,* as they are called elsewhere in Uzbekistan (Kandiyoti 1999: 513) – are an example of the worsening of the former *kolxozchi*s' working conditions. In the switch from kolkhoz to shirkat and then to employment by fermers, ordinary rural labourers have seen their rights diminish and their economic condition gradually worsen.

Despite a degree of continuity in labour relations from the kolkhoz to the shirkat and the fermer enterprise, the way in which agricultural work and control over it are organised has changed. As an employee of an MTP put it: 'There was no unemployment in the kolkhoz. Everybody worked, had to work. If someone was lazy, drunk, or didn't show up, the *sho'ro* came, or the militia. Not so today; nobody would come. But the fermer will chase you away from his land. If you don't work well, the fermer won't give you any of his land'.

On the farms, the difference between related and unrelated workers undoubtedly mattered and underpinned discrimination in labour relations. But it was the young and women who had experienced the greatest worsening of conditions with the demise of the more regulated working conditions of the past. In 2004, while labour relations under fermers were becoming increasingly hierarchical and exploitative, disparities were also developing along culturally embedded understandings of people's roles as defined by age and gender. These additional disparities made it difficult to fully engage with the language of class relations, just as was the case for the 'poor' and 'wealthy' families during the time of the khanate, the Po'staks and Maxsums, described in chapter 2.

*Fermer*s, District Authorities, and Struggles over Crop Growing

Fermers in Khorezm had a special relationship with the district authorities, a relationship characterised by both antagonism and mutual support. The sizes of farms, the social stratification of the fermers, and personal relations all influenced the way in which the relationship was shaped. Whereas small fermers were afraid of the risks and often were pushed into farming, large fermers, who belonged to a privileged and economically capable class of rural notables, actively sought farm leases and a role in agriculture as fermers.

The fact that current reforms have increased the liabilities of land use does not mean that land has altogether become a negative asset. Rather, the new risks and opportunities have been unequally distributed. The crucial aspect is the terms of usufruct negotiated with the authorities. Assuming that other limiting factors do not intervene, those terms strongly pre-determine whether a fermer family is able to make a profit or runs into debt. Even if fermers receive long-term leases, their status is vulnerable because of the importance of tariffs, terms of usufruct, and other vital regulations over which they have no control.

For fermers, cotton is profitable only if the state plan is fulfilled. Many fermers express their desire to grow cotton because of the security it offers: lower profits are compensated for by subsidised inputs, guaranteed prices, and the guaranteed sale of cotton at the state ginneries. In a context in which, according to Rasanayagam (2003: 21), 'the state is no longer conceived of as enveloping the whole society, providing jobs, housing and comprehensive social services', the state still conserves a bit of its former all-round care for the cotton-growing fermer. The problem starts when district authorities push fermers into growing cotton on unsuitable land, threatening the profitability of their farms.

In the 're-peasantisation of society' in Uzbekistan (Zanca 1999), wages lost their value, and land became the substitute asset for generating income. An important indicator of the social and economic status of fermers is their share of land cropped in rice. For this reason, rice is a political issue on which many conflicting interests converge. Rice, unlike cotton, can be directly consumed by households or sold freely for cash in the bazaars, and it has become important for both small producers (who are oriented more towards growing it for subsistence) and large farm holders (who are inclined to grow it as a cash crop). In Khorezm, rice is the most important of the crops that can be locally marketed, bypassing the state retail system, because it can be cultivated on large areas without requiring too many technical means or a large workforce and because of its marketability as the main ingredient in the staple dish *palov*.

According to calculations I made with some fermers, the average profit from one hectare of rice paddy in Khorezm could be 8 to 10 times higher than the average profit obtained from cotton, after expenditures were subtracted, especially if the cotton harvest was as poor as it was in 2003. An approximate figure for 2003–2004 was US$1,000 in net earnings per hectare of rice.[141]

Fermers trying to enhance their rice cultivation may be pushed by either necessity or the desire for profit. For newly established fermers, cash shortage is a major problem, which they try to solve by planting part of their land in rice. Agricultural speculators are attracted by the large profits promised by rice growing. The lack of alternatives for lucrative investment pushes people with capital obtained outside of agriculture to invest in farming, with the idea of specialising in rice.

---

[141] According to a fermer I interviewed later, in 2006 net earnings per hectare of rice had increased by another third over those of 2004.

District authorities try to contain the external investments, on the grounds that they follow only the logic of short-term profit-making and they harm agriculture (cotton cultivation is threatened by the proximity of rice cultivation because the latter's heightened groundwater level affects neighbouring fields). Furthermore, the authorities argue, the land belongs to the state, and the granting of ijara leases entails some sort of responsibility towards collectivity: by taking part in the state plan, fermers contribute to the national welfare. This is the background of a struggle between ambitious fermers and the district authorities.

The struggle became visible in Yangibozor in 2004, the year after de-collectivisation. A ban on rice production fuelled the anger of many rice-growing fermers against the district authorities. The rice prohibition was not placed on subsidiary smallplots (*qo'shimcha tomorqa*) but on large areas that fermers planted in rice for commercial purposes. An August 2004 article in the online newsletter of the Institute for War and Peace Reporting described the situation[142]:

> Uzbekistan's beleaguered farmers are facing new difficulties this summer after the authorities moved rice off the list of 'strategic crops' grown in the republic. This move should have been a turning point for many farmers, who are now free to sell their rice harvest on the open market without having to worry about meeting quotas or accepting the low prices offered by the state. But many claim that local officials are now preventing them from growing rice in favour of cotton – the republic's biggest money-earner and the favoured crop. ...
>
> Two northern regions have been affected more than most – Khorezm and the autonomous republic of Karakalpakistan, which have specialized in rice production for centuries. Figures released by the agriculture and water ministry suggest that these two regions alone were responsible for three-quarters of the 75,500 tonnes of rice Uzbekistan produced in 2003. ...
>
> Kahramon Yuldashev, who heads the grain department at the agriculture and water ministry, told IWPR that the decision to remove rice from the strategic list was a positive move which would benefit farmers. 'Farmers will have more freedom', he argued. 'Now they can grow rice without the control of the state and will no longer be obliged to hand over part of their harvest at government-set prices, as was the case previously'. Yuldashev, who describes the

---

[142] Institute of War and Peace Reporting Staff in Central Asia, 'Uzbek Rice Paddies Endangered', RCA 309, 20 August 2004 (http://iwpr.net/report-news/uzbek-rice-paddies-endangered).

farmers as 'true professionals', told IWPR that he had no information of any threats or damage to the rice crops. But local people insist that these attacks are happening. Villagers claim that more than 140 rice-growing farms in Khorezm and Karakalpakistan have been visited by officials from the local authority and the prosecutor's office and 'persuaded' to abandon their rice crops. Kurbanai Jumaniazova, 40, who has operated a rice farm in the Yangibazar district of Khorezm for years, told IWPR that she had been warned to stop growing rice, and was told that force would be used against her if she refused. Jumaniyazova's farm was later visited by police officers, who allegedly used large tractors to crush the germinating rice shoots. Witnesses spoke of how the farmer threw herself in front of one of the tractors in protest and was dragged to safety by policemen at the last moment. Officials from the Khorezm agricultural department argue that such a hard line is necessary in order to keep cotton production up, on soil that is deteriorating from increased salinity and lack of irrigation water.

What the news article did not mention was the main reason for the interdiction of rice-growing: fermers were diverting the subsidised inputs they obtained for growing cotton to the more profitable cultivation of rice, thus endangering the fulfilment of the district's cotton plan. The availability of water for irrigating rice almost free of charge further subsidised rice-growing. It was not the ecological threat but the threat that free-riding fermers represented to the state plan that frightened the authorities. This point was clearly made in an interview I conducted on the issue of the rice prohibition with a fermer-cum-official in Yangibozor in 2004:

> K: Shall I tell you the main point? You know, the fertilisers for the cotton? Well, these [and] the fuel supply [fermers get through the input supply agencies], instead of using them for the cotton, people put them on rice.
> TT: Does it really happen?
> K: Yes. It is a matter of material interest. For instance, say there are two plots of mine. On one I grow cotton, on the other I grow rice. According to the norms, I have to put one tonne of fertiliser on each hectare of cotton. Instead, I will give only 500 kilograms! My profit from cotton is low, from rice it is high. This is money that goes directly into my pockets. I will give the fertiliser of the one to the other field. Taken from the one and put to the other.
> TT: Does this mean this is the reason [for the ban on rice]. This has nothing to do with ecology?

K: Besides this, there is a link with ecology. Ecology is a very big problem here. From the ecology side, for instance, one has to say the full truth, if we get granted 5 hectares of land, we try to exploit it as if we had 100 hectares to work on. For instance, it is prescribed to grow rice at least one kilometre from the settlements. You shouldn't grow rice close to cotton fields, because then the groundwater rises [and, being salty, damages the cotton]. We don't put these things into practice. Instead, we try not to grow cotton! ... For instance, if at the oblast level the plan is to grow 95,000 hectares of cotton, in the end it falls to 75,000 to 80,000 hectares. The Uzbeks are this kind of people.

The last part of the interview, especially, highlights the way strategic considerations in cropping exist on both sides, fermers and state authorities. Authorities overcharge fermers with the plan, knowing that in the end, production will be lower than initially planned. Fermers cheat on cropping and subsidies, if this enhances their profits. A process of bargaining surrounds the production quotas, and although the situation is clearly asymmetrical, with greater power on the side of the state, it nevertheless offers a different perspective on the belief that fermers are merely the victims of a despotic agricultural policy.

*Fermer*s: Risks, Opportunities, and Family Ties

The establishment of a fermer enterprise is an existential moment for families, because if the state plan is not matched, they will incur debts that they will have to pay with their own assets. This kind of liability is new for dehqon families that become fermers, because even if shirkats accumulated debts on behalf of their shareholders (dehqons) during the period of collective agriculture, households were not directly liable. Understandably, families, especially those with few assets, fear the risks involved. On the other hand, if they meet the state's production targets, fermer households have the opportunity to profit. This opportunity is unavailable to the majority of rural families; even in good years dehqon households make modest incomes compared to successful fermer households. In the abstract, then, becoming a fermer is a desirable thing.[143]

But in Khorezm, the harvests were poor in 2001, 2002, and 2003, while the prices of inputs steadily increased and the procurement price for cotton remained comparatively low. Many of the newly established private farms accumulated debts. For those fermers, the economic opportunities of

---

[143] Everyone I interviewed in my small survey agreed that only those who went into private farming had a future in agriculture.

private farming seemed out of reach, whereas its risks and constraints were immediately perceived. In 2004 the harvest was good, most fermers met the cotton target, and the price paid for cotton at the district ginnery rose considerably. Good profits were made, temporarily easing fermers' worries.

Most newly established fermers perceived decollectivisation as a forced 'up-scaling' of family farming. The fermer has to bear the contradictions of being a state-steered but privately owned and family-managed enterprise. Decollectivisation is an attempt to impose the terms of state farming on families: by becoming fermer enterprises, families approach the features and rules of the shirkats, but on a smaller scale. Decollectivisation affects family farming as a formalisation of the diversified economic practices by which rural producers had adapted to state regulations in agriculture. I illustrate these statements with the story of Maxmud Xursandbekov, the de facto manager (*haqiqiy rahbar*) of the farm 'Xursandbek' in Yangibozor district.

The Xursandbek farm, named in memory of Maxmud's father, was established a year before decollectivisation and then enlarged from 4 to 30 hectares (23.3 of which were arable) in the year the shirkat was dismantled. By the age of 45, Maxmud had worked, after returning to his home village from study at the Institute for Irrigation and Agriculture (TIIAME) in Tashkent, as a Komsomol secretary, a shirkat land measurer (*zemlemer*), a *sel'soviet* chairman (*sho'ro*), and finally at the MTP as an area supervisor for the leasing of tractors to fermers (*MTP uchastka boshlig'i*). Because of his experience in the agricultural production hierarchy, his application for farm land was accepted by the farm establishment commission, which was headed by the then shirkat rais, now the MTP rais. According to law, however, employees of the state administration cannot be owners (*rahbar*) of private farms. Because of this, the farm contract was registered in the name of Maxmud's wife, Ozoda, even though she had a small job in bookkeeping at the MTP.

The Xursandbek fermer enterprise was in the hands not only of Maxmud but also of his three brothers, Jumaniyoz, Bahrom, and G'ulum, who participated in different ways in the management and work of the farm. Besides Ozoda, 10 workers were registered on the farm payroll (*shart-nomaviy ishchilar*). They qualified for pension schemes and could receive salaries from the farm account. Maxmud's extended family, composed of his and his brothers' four separate households, was deeply interrelated in the structure of the farm. Yet despite this integration, only five people were officially 'members' of the farm: Jumaniyoz's wife, Bahrom and his wife, and G'ulum and his wife. Maxmud, although neither the legal owner nor a member, was the recognised head of the farm.

The farm's cropping scheme was decided in the district agricultural department and imposed on Maxmud. In 2003 it was 19.3 hectares of cotton and 4 hectares of wheat. The previous year he had been assigned only 4 hectares of wheat, an easier and much more attractive target. Maxmud did not choose the land he received for cultivation in 2003, most of which was saline because a water drainage channel had previously passed through it. But with 60 points of *ball-bonitet*, the soil quality was good in theory, which meant that 2.9 tonnes of cotton per hectare had to be delivered under the production plan. Given the quality of the land, this figure was unlikely to be attained in an average year. To enhance the soil quality, expensive improvements were necessary.

In the cool spring of 2003, cotton seeds had to be replanted several times, creating additional costs for the farm. The district governor, concerned that the district might not fulfil its plan and also following a directive from Tashkent, imposed the use of a plastic cover (*plyonka*) that protected the sprouts from cold, but this was another expenditure for fermers. Additional tractor work was needed on the Xursandbek farm, but fermers have a limited stock of inputs related to their credit lines for their cotton and wheat contracts. All tractors are old and consume more fuel than is calculated in the norms, so additional tractors and fuel have to be found at times when everyone needs them. In 2003 the available tractors were working night and day. Maxmud's younger brother G'ulum was able to help, because he had purchased a tractor at the shirkat auction and worked as a private tractor driver. But extra fuel and a spare part for the tractor gear had to be bought in the bazaar in Urganch.

When the farm plan was endangered – 'If the harvest is two tonnes per hectare, I will be bankrupt', said Maxmud – he diverted the cash for his workers' salaries to cover current expenditures. The workers were instead paid in wheat at the end of the wheat harvest. When it became evident that the plan would not be matched, Maxmud managed to change the terms of his contract with the cotton ginnery and reduce his cotton order, for which he needed permission from the district department for agriculture. Many of the fermers-cum-officials of the state apparatus of agriculture neglected their farms because their primary task was to coordinate and monitor the fulfilment of the state plan at the shirkat or district level. In exchange, a blind eye would be turned if their farms did not perform well. As one of them, Maxmud had a working day of 16 to 18 hours in the spring, most of them spent coordinating tractors and receiving orders from his rais, for whom he had to be available at any time. Maxmud's employment at the MTP gave him no economic advantage; he spent more on fuel for his Zhiguli automobile than he received in salary, which was two sacks of wheat after the harvest. His motivation for working for the MTP was that it gave him small privileges and some protection from excessively bad terms of trade for his fermer enterprise. For instance, he could get around the order to use a *plyonka,* thus avoiding additional credits.

Maxmud's role as de facto head of the family farm (*haqiqiy rahbar*) included handling paperwork related to banks, input supply facilities, and the district administration. He took strategic decisions about farming and irrigation, and he managed the procurement of inputs, his employees' pay, the marketing of crops, and relations with the local agricultural production hierarchy, which monitored farms throughout all the steps of cotton growing. Typical work for the *rahbar* might be arguing with MTP employees and neighbouring fermers if the irrigation pump upstream failed to work. The job was fully compatible with Maxmud's official occupation. He was the only member of his extended family with a car, without which he could not have effectively managed either the farm or his job at the MTP. The daily management of the workers on the farm was the role of the farm supervisor (*ish yurutuvchi*), a role that existed on every farm of medium size or larger. On Maxmud's farm, the 30-year-old son-in-law (*kuyov*) of Maxmud's elder sister served as supervisor.

Maxmud's extended family perceived the farm as a commonwealth, to which the households contributed as they could. Among the family, some people were affiliated with the farm officially, and others, unofficially. A stratification existed between farm 'members', who were affiliated with the extended family, and employed farm workers, who were not and who did not participate in either earnings or risks in the same way affiliated participants did. Maxmud drew a distinction between farm members (*fermer xo'jalikning a'zolari* or simply *a'zo*s) and simple farm workers (*xizmatchi, ishchi*), although this distinction did not exist in the farm papers. On paper, both groups had the same status as farm employees (*shartnomaviy ishchilar*). On Maxmud's farm there were six 'external' farm employees, all of whom were fathers in their thirties or forties, living in the nearby village, at walking distance from their daily work on the farm plots. Maxmud hired additional seasonal workers (*yo'llanma ishchilar*) for the harvest.

Maxmud's farm illustrates the way fermer enterprises in Khorezm are complexly articulated. Its structure and problems are typical of medium-size cotton and grain farms, which, after decollectivisation, bear the largest portion of state-ordered production. Tractor shortages, cash shortages, uneasy working relations, exposure to environmental threats (and the inability to react to them adequately because of the plan), and exposure to the 'threats of the plan' itself make private farming a burdensome business.

Although all rural households in Uzbekistan are entitled to a subsidiary plot (*qo'shimcha tomorqa*) and land for housing, not all dehqon households engage further in agriculture. Some, although still dehqons, have only loose links with agriculture. Others, whose main family profile is external to agriculture, may 'buy' some land illegally, meaning that they pay for the use of a plot for one growing season in order to make extra cash by growing rice or another cash crop. Of those who do engage in agriculture, some are employed by fermers, and some have become fermers themselves. A mix of different forms of land use is common and corresponds to the varying structure of extended families.

An example is one extended family of 12, composed of two households and four sons, two of whom were married. The family had two tomorqa plots, one for each household, and an arrangement with a fermer to look after a parcel of land (they called this a pudrat contract, although strictly speaking the term *pudrat* referred to a contract between a shirkat and a pudratchi before decollectivisation). This family also had its own fermer enterprise, consisting of one hectare of orchard and some additional land on which to grow rice in the summer, obtained unofficially from another fermer by cash payment. The women of the extended family worked mostly in the cotton fields. Both men and women worked on the tomorqa and in the orchard, and rice was a male domain. Family resources were pooled, and work was done jointly. I interviewed the family patriarch, a retired kolkhoz employee, shortly after the shirkat was privatised. He had decided that his married sons should not quit agriculture (although one of them also worked as a taxi driver), because he believed his family had good prospects for making a living that way.

Some land estates appeared smaller than they really were, a phenomenon in which family links played an important role. Families were strategically mobilised as vehicles to ensure control over land and profitable cropping arrangements. Since decollectivisation, families have adapted rapidly by recomposing social differences already in existence during the kolkhoz period. The scale of farming is an important indicator of this. In the case of Maxmud Xursandbekov's farm – an average case – an extended family composed of four households coped with one farm. In another case, the patriarch of an influential extended family exerted control over several fermer enterprises through his sons and other affiliated and even unrelated straw men (see chapter 6). On paper, the farms controlled by this patriarch were separate fermer enterprises, of modest size, belonging to different persons. But because these persons were brothers, the extended family effectively controlled large portions of land. Put together, as the farms were in fact owned and managed, they represented a large estate. Thus, family linkages helped concentrate land in the hands of a few powerful estate holders who exerted indirect control over their land through different arrangements, mostly based on sharecropping. As a consequence, the de facto relations of land distribution were even more polarised than they appear in the official figures.

In Bo'ston, for instance, two brothers, both with pasts on the steering committee of the former sovkhoz, owned the two largest farms, Buniyod and Jo'ra-Sarvar, at the time of my fieldwork. Consequently, the people of Bo'ston referred to Safarboy buva, the father of the two fermers and the eldest member of his *avlod,* as the largest estate holder in the former shirkat. Although strictly speaking the two farms were separate, and the family patriarch was not directly involved in the everyday issues of farming, as head of the extended family he was perceived to be the owner in a practical sense. His sons conformed to the rules of conduct for a traditional family by seeking their father's advice and instruction in every important decision regarding the family and the use of land. In Safarboy's case, the land estate and kinship network were intertwined.

The pooling of land resources was not exclusively an elite phenomenon. Another example from Bo'ston, involving a less wealthy and influential *avlod* than Safarboy's, was that of the family patriarch Tojiboy buva. His was a large family, comprising 44 members in nine households, which altogether had received 23.4 hectares of farmland. Each of Tojiboy's three sons owned a fermer enterprise, two of which were small cotton farms and the third a small orchard. The family also owned several tomorqa. Although rights to land were individual, Tojiboy's extended family pooled its resources and shared burdens and profits alike. The family did not belong to the upper class of large estate owners, but it displayed patterns of land management similar to those of the larger estate holder, the family of Safarboy.

Fermers in Khorezm manage the burden of farming partly by mobilising their families and partly by devolving it onto their unrelated employees. In this context, the affiliation of some fermers with the state agriculture apparatus is both a burden and a benefit. Legally the fermer is an individual in a lease relationship with the state. In everyday practice the fermer enterprise consists of a family group subordinated to the state apparatus of agricultural production. As extended families have adapted to the difficult circumstances of private farming, distinctions have emerged among roles and tasks on the farms. This diversification is explained by the complex setting in which farms are embedded. The examples I have discussed show how extended families were made to fit into state farming, and how families have coped with this.

Decollectivisation and Kinship

The organisation of agriculture in Uzbekistan draws elements from two domains: the state and families. This twofold embedding can best be illustrated by considering the different scales of activity involved in the production of cotton. In rural settlements, the small-scale work, consisting of heavy manual labour in the fields, is managed by resident peasant families, the gektarchis. They materially handle the daily operations of cotton growing – irrigating, weeding, and so forth – during the growing season, roughly March to August, after which the harvest begins. The large cotton fields are apportioned into smaller units that can be matched by the capacities of the families.

At the same time, on the same fields but at a higher level of the agricultural production hierarchy, local staff and officials organise and execute all necessary operations of cultivation, including the use of tractors, the application of agricultural inputs, the taking of decisions about sowing, irrigation above the plot level, and the management of rural labourers. Higher up yet, at the district level, the apparatus culminates in the *hokim*, the district governor, who is directly responsible for the accomplishment of the state plan and who supervises production by overseeing the district-level staff.

The state-directed framework and the family framework have complementary attributes. Production is organised along lines of family structure while rural families adapt to the framework of state agriculture. This pattern was already well established during the Soviet period. In the kolkhoz, housing and work were organised to suit the exigencies of extended families. The kolkhoz provided its members with large houses, built to hold more than a conjugal family. These always included a garden plot and a shed for cattle in the backyard and had abundant space for storage.

Although sovietisation brought many radical changes to agriculture, the traditional structure of the family as the main unit of agricultural production was preserved in a form not greatly changed, despite initial attempts to diminish its role. Although the pre-existing political structure above the family level was liquidated and replaced by a new one, households were integrated into the productive system without altering their key features. With the post-independence land reforms, this pattern was preserved. Although the legal definitions of land usufruct have been rephrased several times since independence, the rural family has remained the essential unit. A market ethos, consumerist trends, and venality have affected family relations and have strengthened egoistic and individualistic trends that put pressure on family solidarity, but a break-up of the family is not in sight.

On the contrary, post-Soviet agricultural reforms have strengthened the role of kinship in labour relations, livelihoods, and access to land. In this respect my results confirm the findings of other researchers (cf. Kandiyoti 2003a; Kandiyoti and Mandel 1998). In comparison with the situation during the kolkhoz period, rural people in a wide range of cases have become more dependent on their family networks, especially for accessing and managing agricultural land. Kinship already played an important role in the brigades, but rural people are now in a situation in which the family has become the only substitute for services once provided by the kolkhoz. During the Soviet period, better salaries and the introduction of formalised employment began to turn agricultural employees into something more similar to urban workers. With regular wages and work schedules on Soviet farms, people valued agricultural land less highly than they do today. The recent increase in the importance of kinship relations pertains not only to the organisation of farming and daily economic cooperation but also to the more political necessity of preserving or augmenting the family's entitlements to land usufruct.

This phenomenon is not unique to Khorezm but reflects a worldwide trend towards the re-efflorescence of kinship and relatedness in the securing of livelihoods (cf. Brandtstätter 2003; Pardo 1996). It finds its equivalent in many other contexts in the postsocialist rural world in which kinship and, more broadly, 'informally organized social networks' can replace 'the vanishing institutions of the socialist state' (Kaneff and Yalçin-Heckmann 2003: 254). In this respect, Pine's (2002) findings for rural Poland, where kinship relations have become more important in redefining work practices, domains of work, and incomes as rural people feel increasingly excluded from the public domain, correlate with the situation I witnessed in Khorezm.

Together with the growing importance of privately held land for rural livelihoods and of family ties for the successful managing of land, decollectivisation in Khorezm has been accompanied by a strengthening of patriarchal values in society (cf. Finke 2002; Thelen 2003; Verdery 2003). People rely more on kin in a situation in which other institutions have ceased to work and in which participation in the political process from below is not an option. In the government's reform policy, the fermer is the one designated to organise and implement this 're-patriarchalisation', for the new incentive structure makes the extended family the only competitive actor in agriculture. Accordingly, state rhetoric propagates 'a notion of authentic Uzbekness based on a locally shared understanding of kinship duties and masculinity' (Megoran 2008: 30), which fits well with the exigencies and condition of the newly established fermers.

Chapter 6
Rural Elites in Competition

Caught between predictions of Islamist upheaval and the menace of regional fragmentation[144], current understandings of the social dynamics of rural Uzbekistan continue to be trapped by 'discourses of danger' (Thompson and Heathershaw 2005) that make it difficult to recognise what is really happening on the ground. 'Clans' and 'patronage networks' are terms widely used in reference to local political dynamics, and some writers stress that the key political processes are informal and revolve around relations 'between clans and regimes' (Collins 2004: 230). Yet depictions of the underlying social and political configuration have seldom been accurate. Uzbekistan is still portrayed as an essentially static country in which, unlike in most other postsocialist nations, rural areas have been minimally affected by substantial reform. As a result, most literature dealing with rural social developments in Uzbekistan implicitly or explicitly maintains an image of continuity between the kolkhoz of the Soviet period and today's superficially reformed equivalents.

Against these assumptions, I believe recent intensifications of reform have produced a significant shift in political dynamics in rural Uzbekistan. My research in Khorezm shows that assumptions of socio-political immobility and the analytical inferences drawn from those assumptions must be reconsidered in the light of empirical evidence. In this chapter I address significant changes in relations of power and production in Khorezm. A changing scenario is producing a new dynamic that has so far received inadequate attention.

The reforms that led to decollectivisation had the unforeseen consequence of affecting both the official and the implicit rules by which local producers and local elites conducted their interactions within the rigid framework of state agriculture. In the absence of other viable forms of political participation, these rules were the expression of a voice from 'below' in a context in which local communities' administrations and commit-

---

[144] On this, see, for instance, Fathi (2004), Ilkhamov (2001), and Petric (2002: 235ff.).

tees, rather than truly being bodies of self-governance, were made to espouse and follow centrally commanded instructions and policies. Despite Soviet (and post-Soviet) centralism and its institutional animosity towards genuine political participation, power politics never ceased to have a local expression, manifested in local struggles over the distribution of resources and the shaping of community life. In this respect the coping and appropriation strategies adopted by rural communities vis-à-vis the centrally set political frame should be considered a surrogate form of participation, albeit an opaque and indirect one, which the central government has successfully contained until now.

This is the locus of the transformation I address in this chapter by focussing on the way relations between fermers and district authorities have evolved with decollectivisation. Discussing the case of Yangibozor district in greater depth, I argue that local struggles and local forms of coping develop a dynamic that only partially matches the scenarios implied by the centre-periphery narratives of power relations employed to describe sociopolitical ferment at a more general level. In an attempt to understand the centre-periphery struggle in a specific district, I analyse the evolution of the mechanisms through which new inequalities are created and uncover some unsuspected dynamics that create challenges and resistance to state control. Neither 'clans' nor 'solidarity groups' but individuals responding to new sets of constraints and opportunities are at the core of these mechanisms. The terms of the centre-periphery struggle have changed since the period of the cotton scandal, together with forms of patronage at the level of local agriculture.

Centre-Periphery Dynamics in Uzbekistan's Rural Sector

The notion of an all-encompassing state has frequently been invoked to characterise state-society relations in Uzbekistan. Nevertheless, according to Ilkhamov (2004), the dominance of the Uzbek state over society only hides the frailty of its power. The absence of an independent public sphere does not render state power more stable, because 'stability remains based on continuous negotiations between layers of the state that largely leave out society' (Ilkhamov 2004: 162). In such a context, centre-periphery dynamics represent the motor of social and political transformation: they replace state-society relations in the role usually ascribed to them in other modernising countries (as understood, for example, in Migdal 1988). Ilkhamov writes of this centre-periphery dynamic in antagonistic terms. It is characterised by the paradox of a strong centre permanently endangered by the destabilising claims of a locally rooted periphery. His argument is substantiated by what he calls the 'battle for cotton' (Ilkhamov 2004: 162–170), in which an only seemingly strong centre, represented by the top executive power, struggles to fully impose itself on a recalcitrant periphery, represented by regional elites within the state apparatus. The argument is that since the loyalty and acquiescence of the regional elites depends largely on the centre's (in)ability to satisfy their budgetary demands, the latter end up posing a real threat to the centre.

Looked at from a general perspective, the battle for cotton originates in the diverging economic and political interests between the central government, local producers, and regional elites in respect to an agricultural system once built up to serve the demands of the Soviet textile industry (cf. Gleason 1983; Rumer 1989) and now evolving towards newly defined domestic concerns (Kandiyoti 2003a: 227). During the Soviet period the central government implemented a cotton quasi-monoculture in the areas suitable for the crop. Local producers became employees integrated into the kolkhoz system, for which the Soviet government, via local administrators, allocated consistent budget resources. The regional elites gained some discretion over the state budget for agriculture. As I have already described, their technique consisted in over-reporting their cotton production, a device that enabled state enterprises and organisations to obtain larger budget and resource transfers from the central administration. Local elites used these rents to satisfy their needs and those of their clients, and they succeeded – partly by means of the cultural complicity that linked them to their communities – in developing a pervasive system of patronage, ranging from the higher echelons of the state and party system down to the kolkhoz staff and their local constituents (Patnaik 1996: 155–173; Willerton 1992). According to Ilkhamov (2000: 10–11), this legacy of the past lives on in postsocialist relations:

> More often than not, kolkhoz chairmen find patrons among local authorities with whom they enter into mutually beneficial clan-like relationships. This patron-client network pulls in other influential forces, such as heads of law enforcement agencies, bazarkoms (administrators of local bazaars), and even sometimes people from the central government. ... Ordinary rural households are also in some way involved in these patron-client networks. The role of patrons towards them is played by kolkhoz chiefs. From this point of view such kolkhoz-household relationships can be considered a sort of social archetype inherited from the pre-Soviet past.

After independence the government privileged a conservative course of reform, the aim of which was to enhance the condition of a poorly performing and deficit-ridden agricultural economy without loosening its grip over the returns of agriculture. Stability concerns were prioritised over the need to reform the inefficiencies of the state-controlled collectives. Superficially reformed into cooperatives and later into farmers' unions, they maintained their most important characteristics. However, the government was unable to ensure the regular payment of wages and to provide for those social services and investments that in the past had effectively counterbalanced its extractive cotton policy. Budget cuts were aimed at reducing regional elites' capacity to ensure their rents from agriculture (cf. Jones Luong 2002: 131–132). These measures naturally clashed with the consolidated web of interests, collusions, and tactics developed by stakeholders at various levels, which in the past had already led to the so-called cotton scandal, or 'Uzbek affair'.

As a consequence, post-independence agropolicy created resentment and hardship among primary producers. In order to compensate for its incapacity to maintain past living standards in rural areas starting from the 1990s, the government enlarged the share of land devoted to household use and reduced the overall share of land grown in state-ordered crops (Lerman 1998: 168; also see chapter 4), but with little success. Local elites, deprived of their share of rents from the business in cotton and other state crops, were compelled to seek alternative sources of income, which they found at the expense of the rural population. The result was to further penalise the primary producers, who were now forced to endure a 'double burden imposed on them by the central government and their local bosses' (Ilkhamov 2004: 168). In the 'colonial' (Akiner 1998) context of late socialist Uzbekistan, local elites were appreciated as supportive patrons of their communities. Their capacity to divert and allocate resources to their patronage networks could even find a moral legitimacy by figuring as a passive form of resistance against the centrally imposed cotton programme. In this sense, the local elites demonstrated their loyalty by 'stealing' from Moscow to 'give back' to the communities. In the context of independent Uzbekistan, by imposing an additional burden on the communities, they lost the legitimating 'Robin Hood bonus'. The moral support rural elites enjoyed in the past now fades and gives way to a post-independence legitimacy crisis (Fierman 1997: 393).

As Gleason (1991) acknowledged, during the years of the cotton scandal the position of the rural elites was ambivalent, moving back and forth between the interests of the political centre and those of their local constituents, depending on the situation. According to Ilkhamov (2004), this ambivalence has continued. The regional elites share with local producers an interest in expanding the sector of agriculture not devoted to cotton, because this is the sector they can control autonomously and which allows them to earn the material resources no longer available through the governmental budget. The centre's reaction against this strategy is to augment its pressure in the form of periodical staff reshuffles and intensified controls on production. This antagonistic relationship between centre and periphery produces what Ilkhamov has called the 'battle for cotton'.

Although this reconstruction of the political and economic dynamics surrounding the cotton sector is plausible in general terms, little is known about how these dynamics concretely affect local elites and producers in rural areas. In the following section I transpose this model to my case study district in Khorezm, in order to see how far the paradox created by centre-periphery dynamics is reproduced at a lower level of the agricultural production hierarchy and which particular form it assumes.

The Battle for Cotton Transposed to Yangibozor

Ilkhamov's model of centre-periphery relations in Uzbekistan is based on his understanding of the dynamics between the top-level state hierarchy and regional elites. Seen from the cotton-growing district in which I conducted my fieldwork, the scenario described as the battle for cotton is not immediately recognisable.

Ilkhamov's model emphasises the rivalry between regional elites and the central government, but in Yangibozor the district establishment seemed to be genuinely loyal and did not represent a challenge to the government. The district governor (*hokim*), for example, had been installed in order to carry out the policy of decollectivisation, and he was later praised for his successful accomplishment by the regional hokim for Khorezm. At the time of my fieldwork the district hokim was a machine engineer, originally from neighbouring Gurlan district, who upon his arrival in Yangibozor served as city hokim in the nearby regional capital, Urganch, where he still resided. To residents of Yangibozor who were critical of him, his 'foreign' origin was another reason he supposedly was 'more after the interests of the state than after those of the district'.

Also, local producers in Yangibozor were much less passive than those portrayed in Ilkhamov's account. Yangibozor has in the past mostly met the cotton production targets set by the government, although at the cost of many sacrifices and hardships for the direct producers. Under the *shirkat*s the opportunity to make profits from rice, which district authorities tried to limit in order not to endanger cotton production, as well as to keep it under their exclusive control, was very attractive to local producers. Now, with land directly available to *fermer*s, its appeal has only increased. Attempts by the hokim and the other district-level staff to keep the rice sector under their control has not gone unchallenged locally.

Representing post-reform Yangibozor as a scene of conflict or social unrest would be mistaken. Yet behind a harmonious façade, relations between district authorities and recently established fermers were a continuous source of tension rooted in the latter's determination to manage their farms independently and expressed in locally circumscribed phenomena of resistance against the authorities (cf. Wall 2007). In June 2004, one year after the land of the former collective farm Amudaryo was entirely transferred to fermers, these tensions poured out in an exceptional, locally circumscribed but emblematic uprising against the authorities, particularly against the district hokim. A casual spectator described the event to me in these words:

> Yesterday around 10 or 11 in the morning in Amudaryo people gathered on the bridge and stopped traffic, approximately 300 people, most of them *dehqon*s, about 50 of them fermers, the rest people working for the fermers, all were from Amudaryo. ... To calm down the crowd, who were protesting because rice growing had been banned, even A. [head of the regional agricultural department] had to come from Urganch to support the hokim. The people were shouting at the hokim. An angry woman was saying, 'We have already paid for the tractors, bought the seeds, everything is ready, why do you tell us only now to stop?' Then A. had to speak [to explain the reasons for the ban] to calm down the people while the hokim sat aside. In the end the authorities promised that rice would be grown again. But after the people went home the authorities took back their promise and stuck to their 'no' to rice. Now there is no electricity in Amudaryo; the hokim today stopped the supply [so no one could run electric pumps to irrigate the rice fields]. Without water, in a few days the rice paddies will dry up and all be gone.

Although relations between district authorities and producers only rarely deteriorate so overtly, the episode is telling of the types of constraints local producers face and of the asymmetrical relations of power between them and the state authorities. Moreover, it portrays the top-down implementation of policies as far more complex and contested than is usually recognised, and it points at some of the means of coercion used. In this account, local producers speak up against the hokim through the voice of an angry peasant woman, erroneously suggesting that local producers constitute a clear front against the local establishment. Rural producers, as I discussed in chapter 5, are not a homogeneous class or even a 'we-group' in Elwert's minimalist sense (Elwert 1989). In the aftermath of decollectivisation, however, such tensions were the by-products of a new social differentiation with a still uncertain outcome. The reason for the protesters' anger, the interdiction of rice cultivation, brings us closer to the heart of the matter: the possibility and modalities for making profits in agriculture in the setting of an ambiguous reform that creates more space for entrepreneurship while maintaining strong constraints on entrepreneurial freedom. Transposed to a lower rung of the agricultural command hierarchy, the *rayon-tuman* level, Ilkhamov's centre-periphery battlefield scenario unfolds here between the district authorities, represented by the hokim, and the (newly established) producers who do not wish to comply with their orders and abuses.

Reproducing Dependency Patterns

Fermers welcomed the end of the shirkats. For them, decollectivisation, although it came in the form of conditional land leases, entailed legal claims on land and the end of their dependency on the former kolkhoz bosses. In practice, conditions did not change so radically, because by law fermers still depended on a network of nominally privatised but de facto state-controlled input and marketing services that made them easily fall prey to state officials. It was precisely the inadequacy of conditions surrounding their farm enterprises that rendered fermers so vulnerable to unpredictable factors of risk. The dependency patterns established under collective agriculture survived decollectivisation and were transposed onto the new scenario.

In line with Christophe's (2005) analysis of the Georgian state, in rural Uzbekistan the institutionalised creation of insecurity, whether through structural system shortcomings or outright intimidation of producers, appears to be a calculated means of state-building. Uzbekistan's regime of agrarian transition is reminiscent of contexts for which Verdery (1999) developed the notion of 'fuzzy property'. At the end of decollectivisation in Romania, she observed that the external conditions in which legal ownership titles were embedded were much more important than the titles in a narrow sense, because those conditions, over which producers had limited influence, defined what ownership rights ultimately entailed.[145]

Uzbekistan's situation is similar. Often misunderstood in Western-minded reform agendas, the surrounding conditions, including the persistence of pseudo-collective structures, the social embedding of work, the

---

[145] Verdery (1999: 64–65) wrote: 'Fuzzy property, in the examples given here, consists of complexly overlapping use and revenue rights lodged in external conditions that give the holders of those rights incomplete powers for exercising them. The external conditions include such "legacies of socialism" as a systemic bias against individual ownership and in favour of state or quasi-collective forms. For something more closely approaching exclusive individual proprietorship to emerge would require not so much clearer legal specification of who has what rights – these rights are fairly clear – but modifications in the surrounding economy that would permit individuals to acquire the means of cultivation affordably and to dispose of their product profitably while outcompeting quasi-collective associational forms'.

necessary web of personal relations, the quality of infrastructure, and the availability of markets, have become the critical pressure points in decollectivised production. These factors, rather than secure ownership titles, end up being what really matters to producers. As Rasanayagam (2002b: 18) put it, in Uzbekistan rights over land are 'likely to be [even] more fuzzy'. This is because private property in land still has not been introduced, causing actors to have more asymmetrical influence on access to, use of, and control over land than in the context Verdery analysed. Legislation also lacks clarity and gives rise to interpretative difficulties, which authorities often use as a last resort to make fermers comply with their orders.[146]

In Yangibozor, the 'fuzziness' of fermers' rights over land is a result of three main factors. The first is a technical one: a shortage of equipment and machinery for fermers to use. The 11 MTPs in Yangibozor that hold the still unprivatised machinery of the shirkats, although officially independent service enterprises, in reality are follow-ups of the kolkhozes, controlled by the district hokim, who appoints the managers (*raislar*). The MTPs lack sufficient machinery to meet demand, so they act as bottlenecks in production and help re-create relations of dependency for fermers.

The second factor is that the district authorities manipulate the MTPs and other supposed service organisations, such as the Fermer and Dehqon Association (FDA), for their own ends. As a consequence, the FDA, formally established as a republic-level organisation with regional and district-level branches and the declared aim of supporting the development of private farms, in fact sidelines the district department of agriculture in coordinating the state crop planning.

The third factor is the uncertainty that the legal-administrative realm of sanction creates for fermers. The MTPs are the pretext for maintaining the executive structure of the former kolkhozes and shirkats. Each shirkat, like the kolkhoz before it, had a rais and his staff: a deputy rais, an agronomist, an engineer, a land measurer, and so forth. After the disbandment of the shirkats, these people became affiliated with the MTPs, where they continue to perform the tasks they did in the past.[147] The MTPs, the banks, the agencies that supply agricultural inputs, and the district department for agriculture all form indirect levers through which the district hokim can make fermers compliant to his will. In a certain sense the hokim has taken the place of the kolkhoz rais. Wealthy fermers with their own equipment are less

---

[146] An example of inconsistencies in laws can be found in the Land Code, 30 April 1998, paragraphs 5 and 32, which give contradicting reasons for farm closure (information drawn from an interview with an official of the Fermer and Dehqon Association in Yangibozor).

[147] Officially these officers are employed by newly created, district-level organisations. In practice they are subordinated to the MTP rais, who can replace them at will.

dependent on the MTPs, but they can still be intimidated through the periodic checks and controls of the prosecutor's office and the militia.

Plate 13. Entrance to the former kolkhoz 'O'zbekiston', later a *shirkat* and then an MTP, in Yangibozor district. September 2006.

Officials of the district agricultural department use these instruments to impose centrally determined cropping arrangements on fermers, when according to law fermers should be able to decide freely what to grow on a significant share of their leased land.[148] Because the imposed cropping arrangements are key to the profitability of the agricultural enterprises, they represent the political lever that empowers district officials in their dealings with fermers. Although legislation and norms prescribe equitable treatment of farms, district officials have a margin of discretion, which they use to turn cropping arrangements into a commodity. In the past this was a prerogative of the kolkhoz bosses, who would 'assert private claims over land as a valuable resource under their control from which to obtain personal profit through the illegal transfer of rights over it' (Rasanayagam 2002b: 19). Now the district hokim stands out as the only official entitled to grant land leases

---

[148] At the time of my fieldwork the law stated that only 30 per cent of the land of the farm leaseholds had to be grown in cotton, so in theory the residual land could be cultivated freely.

to fermers, and he struggles to keep land transactions under his control. The logic of the transaction is that of selling cropping patterns in exchange for cash or a share of the harvest. Cropping arrangements are negotiated individually and differ in the share allocated to each crop from farm to farm. Agreements that include rice and other non-state-ordered crops are confidential and are settled secretively. State officials dealing with these data keep them secret because a comparison of the cropping schemes of different farms in a district would reveal that the state plan was applied to fermers unevenly.

To some extent, these opportunities for profit are used as rewards to compensate people for their loyalty and their role in the command chain in a context in which, after the end of the shirkats, local officials have maintained their obligations towards the command hierarchy but have lost the direct instruments with which to ensure the fulfilment of crop production targets. By law, it is not possible to be head of a farm and at the same time an officer in the state administration. Yet indirect control over farms is common. In Yangibozor in 2004, all MTP managers and other high-ranking district officials made their livings from their own large farm enterprises. In line with past practices, privileged land deals were granted to particular fermers who held roles in the local state apparatus, as a means to ensure that the plan on which the political survival of the local authorities depended was fulfilled (Thurman 1999).

After decollectivisation, access to land leasing and, especially, to profitable terms of land usufruct became a substitute for the adequate salaries that state employees no longer received. When I visited Yangibozor, every MTP rais was also the owner of a large farm with a profitable cropping scheme. Many of the raislar I interviewed said they would not mind giving up their position and concentrating on their own farm and interests, but this would displease their hokim, who needed them to monitor the growing of crops and collection of harvests. With monthly salaries equivalent to US$20–30 – not even enough to cover the fuel expenditures necessary to accomplish their job – this was understandable. Through farming, if profitable arrangements were made and cultivation was done properly, they could earn tens of thousands of dollars. The opinion of Pulat aka, the rais of a shirkat in neighbouring Shovot district at the time I interviewed him, well illustrates this effect of decollectivisation:

> TT: What will you do when the shirkat is dismantled?
> Pulat: I will become a fermer, what else?
> TT: Is it not worse to be a simple fermer, compared to being a rais?
> Pulat: Not necessarily – it may even be better. With 50 hectares of land, if you work skilfully, you can earn 100 million. But don't write it down in your notes; this is not necessary to your work.

Beyond the fulfilment of the plan, the concern of the district heads seems to be control over the sources of wealth in the district. Because of their assumed wealth, particularly well-to-do fermers are targeted with continuous requests for payments, contributions, and 'one-time taxations' (*chakana nalog*). The last can be requested by district officials on the pretext of any public concern, such as the need for donations for the poor, the renovation of public buildings, public celebrations, and other expenditures for which the district is short of money. In Yangibozor, fermers annually donate a percentage of the harvest of the state wheat crop and of the rice harvest to an ad hoc philanthropic foundation for services such as the renovation of the district hospital or the stadium. These requests signal that the district authorities want to make fermers aware of their social responsibility towards their communities, but fermers complain about their coerciveness and lack of transparency. In a discussion with Egambergan, a wealthy rice fermer from Amudaryo, it was evident that the pressure exerted by the district authorities, even if unspoken, had a strong element of intimidation:

> TT: What will happen if you don't give money, if you don't help the hospital, if you resist *chakana nalog?* These things have no legal basis.
>
> Egambergan: You have to give. These are people with whom you cannot play: they will call the militia, they will find something, for example they will beat up someone and say you did it, or they will find narcotics on you, they will send you to jail, or take your land. You have to give. I am also thinking about whether to stay or leave [agriculture].

Obedience to the requests of the hierarchy, conformity to orders, and avoidance of legal disputes are preconditions for successful farming at least as important as the 'hardware' of agriculture: good land, sufficient water, a workforce, knowledge, and equipment.

Old Established versus Newly Emerging Rural Elites

If in Yangibozor's battle for cotton the centre-periphery antagonism takes only the form of a tug-of-war between local authorities and non-complying fermers, then so far this story has revealed nothing paradoxical about the centre's strength. But the agents of the state are less pervasively powerful, and fermers are less weak and defenceless, than so far portrayed. The fuzziness that I invoked as a reason for fermers' submission to the district authorities can even turn out to be a boon for fermers. As I discussed in chapter 5, the subsidies that enable the cotton ginneries to buy fermers' cotton cheaply also make one hectare of rice paddy, on average, 8 to 10 times more profitable and less labour intensive than cotton. For the authorities this creates a need to ensure that subsidised inputs, especially fertilisers, do not end up on the wrong fields, a task that is beyond the capacity of the district *hokimiyat*. Every year, rice bans proclaimed at the beginning of the season, often on the pretext of (real or pretended) water shortages, are reconsidered later in the season and finally are drastically revised. Fermers plan their field work and cropping strategies accordingly. The eventuality of interference by the district authorities or the militia is a calculated risk that fermers try to mitigate with the strategies and resources they have personally available. This sort of arm wrestling between planners and farm managers is not restricted to cropping decisions but extends to the uses and yields of farm land.

In this context I heard the term *sponsor* applied to a form of indirect control of a farm, a usage that bears the marks of a patronage relationship. In an interview, Komilbek, a fermer and university teacher, explained:

> There is a person who is officially in charge [the official leasehold owner], and there is somebody who is really responsible for the fermer, who is really the head [of the farm]. For instance, if you don't have enough fertiliser, or if you can't rely on the work of your own tractor, what will you do? Who will help? ... So for instance there we have K.H., owner on paper of a farm of almost 70 hectares, but M.H., secretary [*muovin*] of the MTP, is the 'sponsor', or real owner. He uses the tractors of the MTP as he pleases on the farm, and so he makes a very good deal. Normally, the machinery of the MTP is very unreliable: you cannot be sure of getting it in time, or sometimes of getting it at all. So if you have a good connection with the MTP, you can rely on it; otherwise you have to use your own! So you have a person who is legally registered and a person who controls and factually owns the farm.

Informed by a literature that postulated the centrality of clan-like patronage networks in relations between centre and periphery (Collins 2004; Schatz 2004), I expected to find such structures as local carriers of resistance in Khorezm or as vehicles of some sort of political counter-claim. That expectation was disappointed. Instead, hidden farm 'sponsoring' emerged as a proactive, profit-seeking strategy that, although strictly speaking illegal, was widely tolerated. The law prohibited land sub-leasing, and land transactions had to be officially channelled through the district heads. In Khorezm, hidden farm sponsoring was a widespread way of circumventing these restrictions. External farm sponsoring was an elite phenomenon, mostly involving large areas of farm land that were sub-leased under sharecropping arrangements or 'sold' unofficially to the tillers in a deal mutually advantageous to the contractors. As a patron-client relationship, it was eminently business orientated and thus different from relationships based on ties of family, 'clan', or other sorts of solidarity arising from shared identity. Fermer enterprises, as joint enterprises of extended families, always featured a degree of informality, but this was different from the informality of sponsoring, which always featured a degree of impersonality. Moreover, a sponsor was exposed to potential legal retaliation, which he had be able to avoid or resist, so sponsoring necessitated political skills in the sense that economic success emanated directly from the sponsor's power.

Baxtiyor, a language teacher from Yangibozor who occasionally illegally sub-leased land from a fermer in order to sharecrop rice, distinguished two types of fermer enterprises according to their management structure: 'In the first type the owner sits in Urganch, never comes to see the farm, and takes the profit. Owners of this kind can be, for instance, *prokuror, militsiya, tashkilot* [state organisation]. The second type is the ordinary person. As soon as he becomes a farmer, he tries to manage everything himself in order to save money'. In Yangibozor, in the aftermath of decollectivisation, most fermers had just started their enterprises and faced strained or even difficult conditions. Big farms were exceptional, and their presence often indicated the existence of a sponsor, because of the greater capital inputs they required. In a situation in which conditions for overall profitability depended heavily on factors external to the farm, formal ownership was less important than de facto control over the farm as a generator of profit.

In Yangibozor, being a fermer had acquired increasingly political overtones in the sense that a successful fermer needed the 'strength' (*kuch*), capacity, and capital to impose himself on the endemically adverse circumstances of the local agricultural setting. With the end of the shirkats, land became more easily accessible to wealthy people who had no links to the former collectives, because it was now redistributed through the district centre. Bypassing the MTP chairmen, emerging political entrepreneurs (sponsors) could more easily access land and sidestep the former kolkhoz-shirkat administration. In the past, these kinds of arrangements required the use of the land of a collective farm and the mediation of the shirkat chairman. Although asymmetrical power relations persist, today these new actors are determined to achieve lucrative cropping arrangements and are unwilling to retreat in the face of the paternalistic reprisals of the district authorities.

In the perception of Egambergan, the wealthy rice fermer from Amu-daryo, today's fermers are 'reverting to the time of the beys' as they submit to their bosses (*yoshulli*), who pretend to rule as the beys did in the past. The conflictual relationship between authorities and their challengers is expressed through the question of who exerts 'fatherly authority' (*otalik qil-*

moq) over fermers and their resources.[149] In the local dynamic of competition, sponsors' claims to paternalistic control of farm yields threatens the hokim's exclusive primacy in the district. Interestingly, by referring to *otalik,* the terminology of this competition invokes the language of traditional politics with reference to eldership. But the revival of traditional political terminology does not imply that a return to traditional political structures is under way, just as today's sponsors are not clan elders.

Fortuitously, I came across an official document prepared by the district-level department for agriculture for the Yangibozor prosecutor's office that was revealing of the struggles between sponsors and state authorities. The document listed 118 farms – at the time approximately one-tenth of all farms of the district – that were allegedly controlled by external sponsors and needed to be checked for irregularities. All the farms listed were well above average size. The document included the names of the registered titular lease holders and, in separate columns, the suspected sponsors and their professional affiliations. According to Komilbek, the fermer and university teacher I mentioned earlier, and to two former officials in the district department whom I interviewed, the list had been compiled deliberately to cause trouble for the hokim's potential enemies, in order to make them comply with his requests. Komilbek told me:

> The hokim orders the preparation of such lists by the agricultural department for the *prokuratura.* ... He orders it because these are people who have another protector [sponsor] and don't pay him [the hokim]. Everybody pays, but these people don't pay so much. For instance, here we have a fermer controlled by a guy from the militia. The hokim says to him, 'Give me two tonnes of wheat' [at the end of the harvest]. The man from the militia says, 'Don't give it to him'. There is competition on this. Therefore these people are rivals of the hokim, and the hokim ordered the list for the *prokuratura* in order to give them trouble.

In Ilkhamov's national battle for cotton, the pressure exerted by the centre on local elites had 'resulted in shirkats being frequently visited by representatives from the prosecutor's office and the state militia who have been charged with enforcing cotton and wheat land quotas and preventing farms from growing crops that are more lucrative – for them as well as for

---

[149] *Otalik* derives from *ota,* meaning 'father' but also 'ancestor' or 'progenitor'. *Otalik qilmoq* literally means 'to assume the role of a father', 'to educate' (see *O'zbek Tilining Izoxli Lug'ati* 1981). Historically, the term *otalik* was used to mean 'a guardian and tutor of a young prince and, in this capacity, an actual governor of his appanage' (Bregel 2004). Later the term came to refer to the Uzbek tribal leader of the *beg* or bey sort or to the advisor of a khan. Today the term has lost this meaning in everyday language and is used to denote paternalism, or a fatherly role, over someone or something.

regional elites' (Ilkhamov 2004: 168). Seen from the district perspective, such checks and controls often follow local rationales that can be entirely different from those from which they nominally originate. They can begin with personal rivalry or be manipulated for purposes such as intimidation, blackmail, and (as in this case) the hokim's attempt to affirm his 'paternity' over the whole district. He knows that if he does not, then sooner or later his authority will be challenged by the individualism of the emerging fermers.

Changing Patterns of Patronage

The new prospects for achieving wealth through agriculture have attracted people who in Yangibozor were called 'new farmers' (*yangi fermerlar*). These newcomers to agriculture are people who, enriched by capital acquired in other sectors (militia, business, state organisations) and struggling to find sectors in which to invest their money securely and lucratively, enter agriculture with the idea of obtaining considerable material gains through informal deals on profitable cropping schemes. In the words of an MTP land surveyor with whom I worked, these new actors 'see the growing crops as if they were banknotes' but otherwise are not knowledgeable about agriculture. They mobilise their social and political capital to ensure profitable farming without getting directly involved in it. Typically, the structure of the farms of these newcomers is reminiscent of absentee landlordism, in that the farm owners exercise external control while outsourcing the management of the farm to local workers on the basis of sharecropping agreements.

The emergence of the 'new farmers' can be traced to a more general phenomenon. Reversing a trend that was strong during the Soviet period, towns in Uzbekistan are now gaining advantage over villages.[150] Before independence, villages had become reservoirs of political, cultural, and economic autonomy, because kolkhozes could be run as relatively autonomous fiefdoms in which external interference was slight and was mediated by the central hierarchy, embodied by the rais. After independence, towns regained control over villages by refining the mechanisms for extracting the surpluses produced in and around agriculture. This resource extraction is now easier and more efficient than before. As a result, cash shortages afflict rural households at a time when cash can no longer be easily replaced by 'manipulable resources' (Humphrey 1998: 195ff.). One land measurer in Yangibozor said that villagers now 'sell their cows to satisfy the city's appetite'. Villages' dependence on towns has increased; sanitation, higher education, bazaars, and all the administrative offices are in towns. The

---

[150] This appears to be a general postsocialist pattern; see Hann (2003: 39).

'kolkhoz society' (Roy 1999) has lost its economic basis, and its decline is evidenced by the decline in the status of the rais, its symbol of authority.

Big fermers, newcomers to agriculture, and village outsiders who obtain land by paying bribes and then grow rice mostly have bases in town: they are businessmen, doctors, militiamen, and employees of administrative departments or of the hokimiyat. Even in agriculture, many among the new bosses are construction engineers, architects, or accountants with experience in administration. They have supplanted a generation of raislar educated as agronomists, who now find themselves managers of MTPs without adequate educational background.

Table 3. Changing forms and contents of the local political game in Yangibozor district.

Variable	During Collective Agriculture	After Decollectivisation
Main contested resource	Budget allocation	Freely marketable crops (rice)
Competing actors	Kolkhoz staff vs. external controllers	*Fermer*s (old producers and newcomers) vs. local authorities
Form of resistance	Under-reporting of production	Diversified struggle for political control over farming
Modality of resistance	Monolithic patronage pattern	Hidden practice of farm sponsoring

In Yangibozor, against all appearances of 'weak reform', decollectivisation has brought about significant transformations in both the substance and the forms of the ongoing struggle between central assertions of power and local resistance (table 3). Among the actors able to influence resource flows locally, a redirection of the local political game has taken place in the shift from a mostly subsidised to a mostly taxed agriculture. The rationale of political action has moved away from control over budget resources accessed by actors by virtue of their affiliation with the quasi-state collectives, which those actors could allocate to their constituents, towards the possibility of producing for and gaining access to markets. Under the kolkhoz system, the source of competition was control over the budget and its redistribution, which 'outside' control aimed at keeping on predetermined tracks. Today, local entrepreneurs seek the opportunity to make profits and face constraints from local authorities, who try to keep the emerging market dynamics under their oversight.

The shift is from a political game governed by a redistributive logic to one governed by a more market-oriented logic. This shift towards the market, although still imperfect, has wide-ranging consequences. It modifies the economic basis of patronage relations, which have lost much of their past appeal and therefore need a new mode of functioning to gain consensus. The monolithic patronage system centred on the kolkhoz and its network has been replaced by a more diffuse practice of indirect control (more entrepreneurial and competitive) over the means of production, which finds an important expression in the practice of farm sponsoring. While in the national battle for cotton 'rebellion takes place in the form of underreporting of resource use (such as land) and hiding a share of locally acquired wealth from the centre's strict fiscal accounting' (Ilkhamov 2004: 169), in today's Yangibozor rebellion patterns are extended to a diversified struggle over the political conditions surrounding farming.

Local Political Entrepreneurs

As I described in chapter 5, in the aftermath of decollectivisation the MTP Xalqobod counted 134 fermer enterprises. A man called Bozorboy, who had a 10-hectare farm in the MTP, was representative of a class of small fermers whose new status seems to entail more liabilities than opportunities. Since 2000, the year his farm was established, Bozorboy had never managed to fulfil his production target for cotton. When I first met him, he complained that, given the high salinity of his soil, the district authorities had set the production plan for his land unrealistically. Even in the unlikely event of plan fulfilment, his profits would be low; according to him, if the authorities let him grow one hectare of vegetables, he would earn almost three times as much as he could with 10 hectares of cotton, a net amount of US$1,000 at harvest in 2003. When I asked how he saw himself as a fermer in 10 years' time, Bozorboy answered, 'Not in 10 years, in 2 years I will be bankrupt and lose the land. They don't care about bringing people to bankruptcy. It is like a feudal system!' He would later change this attitude dramatically.

The son of local notables (his grandfather was rais of a forest compound, and his father, a militia officer and World War II veteran), Bozorboy, after studying law in Tashkent, had returned to his native kolkhoz to marry a half-Tatar woman. He served in the militia until the early 1990s, when, for reasons that remained unclear to me, he was discharged after a trial. After that he ended up working as legal advisor in the district branch of the Fermer and Dehqon Association (FDA). Additionally, for an unofficial payment he acquired a land lease under his wife's name in 2000 and established a cotton and grain farm, which he named after his father. He made a sharecropping agreement with a former leader of a work brigade on the kolkhoz who lived close to his farm plots. Bozorboy's eldest son, aged 22, worked there as farm supervisor. Asked why he got involved in agriculture, he replied that it was the lack of alternatives that made him take the risk of private farming.

In 2003 Bozorboy was undoubtedly a low-profile fermer in a difficult situation. Either because his land was really unsuitable for cotton or because the sharecropper's family did not put enough work into his fields, the farm was giving poor results. His situation as an FDA employee was also precarious. Because of cash shortages in the district, the yearly FDA salary amounted to two sacks of wheat, and most fermers in the district were not paying their membership fees. The high turnover of appointed managers, former raislar and high district officials, originated in the lack of adequate returns for their engagement and reflected the FDA's weakness. Since my arrival in Yangibozor the FDA office had moved three times, and before 2006 the district branches were combined into a single, regional-level office. For Bozorboy the FDA's only appeal was its being a platform that enabled him to deal with documents of the state administration and to have an official pretext to relate to the agricultural hierarchy. Authorities knew about the sharecropping agreement through which Bozorboy's farming was conducted. Strictly speaking, it was illegal, but a degree of tolerance existed for farms that fulfilled the cotton plan. For this reason, with harvest shortages in 2001 and 2002, Bozorboy had to find ways to 'adjust' the production plan of his farm with the district department officials and buy cotton undercover to deliver back to the cotton ginnery, in order to avoid sanctions. To cover his expenditures, Bozorboy, who had 15 cows before starting his farm, sold 3 cows each year in the city bazaar. In Yangibozor this represented a remarkable capital asset.

Things started to improve in 2004, thanks to Bozorboy's links with the district department of agriculture. Short of personnel trained in jurisprudence, and with an augmented need for legal expertise derived from the sudden increase in the number of fermer enterprises, the department increasingly needed Bozorboy's skills. He helped in the preparation of trials in which fermers were involved, prepared dossiers on farms that had legal disputes with administrative bodies, and followed district staff in their rounds in the former kolkhozes of the district. In his job Bozorboy was continuously involved in the juridical counselling of fermers, but in contrast to the ideal mission of the FDA, his primary concern turned slowly to safeguarding the hokim's interests. 'What the hokim orders, that is my job [*hokim buyurgan shu ishni man bajaraman*]', he said. During farm inspections, a camera, lists of farms with details about their balance sheets and business plans, and correspondence with the court and the prosecutor's office figured among Bozorboy's habitual 'work instruments', testimony to his acquaintance with a methodology of local governance based on a distorted understanding of law and public institutions. Spying for the hokim, looking at the real conditions of farms in the district, he gained the hokim's favour, a precondition for later rewards and success.

In late spring 2004, after a meeting with the deputy hokim, Bozorboy managed to turn the cropping specialisation of his farm from cotton and grain to poultry, paving the way for the suspension of the mandatory state quota for cotton. Thanks to this subterfuge, which the law permits when the administration has certified the unsuitability of a given plot of land for cotton, Bozorboy acquired the possibility of legally turning his cropping scheme into 10 hectares of rice. In 2004 he became one among the handful of fermers in the district who were legally exempted from the rice ban. In the end, he informally turned over an agreed part of his rice harvest to his district bosses, keeping a good profit for himself.

Bozorboy managed to change his situation and turn the poor initial conditions of his farm into an asset, thus realising what in theory is legally possible but which is inaccessible to most fermers: converting cotton land into free cash-cropping land. What were the conditions for his success? Various factors seem to have been important: the plasticity of the law – the arbitrariness of legal processes –which a small caste of insiders can easily manipulate for their own purposes; his wealth of contextual knowledge about many fermers and state officials, which made Bozorboy a useful asset for the power holders in the district; and his knowledge of the rules and tricks of a non-transparent and overregulated agriculture, which made the privilege of state crop exemption legally achievable. Bozorboy's story is indicative of the current transformation in Uzbekistan's agricultural world. In a context in which the market is not yet totally liberalised and the planned economy not totally overcome, he represents the transitory model of agency best adapted to the scenario emerging in Yangibozor: a new market of political protection around the conditions of farming, in which, besides capital, agricultural knowledge, and access to markets, successful entrepreneurial skills have to take power into account.

Equally distant from the centre and the periphery, Bozorboy resembles neither the 'big man' of the political anthropologists (Godelier and Strathern 1991; Sahlins 1963: 285–303), for whom in Uzbekistan, after the decline of the rais, there seems to be no real replacement below the level of the hokim, nor the colonial broker, with that figure's dual allegiances and double burdens towards the state and his community.[151] Unlike the colonial broker,

---

[151] An attempt to explore the role of the local official in a reform socialist setting of Central Asia has been undertaken by Béller-Hann and Hann (1999), who point out the analogy with the African colonial broker.

Bozorboy did not face a dilemma of loyalty but rather strategically switched between conformity to and subversion of the rules of officialdom, a strategy he saw as legitimated by the perceived insecurity of his personal future. The figure to which he probably comes closest is that of Weber's political entrepreneur (cf. Weber 1972 [1921–1922]: 840–860). As an individually driven manager and manipulator of the political conditions surrounding farming, Bozorboy had everything to gain (or to lose) from politics.[152]

From Cotton Scandal to Battle for Cotton

A few months before the end of my fieldwork I received a newspaper article from an employee of the district agricultural department, the *boshqarma*, with whom I had become acquainted during meetings in Yangibozor. The article, titled 'It's your time and use it completely' – meaning 'It's your turn, so take advantage of it' – had appeared on 4 June 2004 in the newspaper *Qishloq hayoti* [Village life], published in Urganch. It reported on the misdeeds of a former rais of the Yangibozor FDA and one of his staff members, both accused of embezzling public money. I present here a shortened, freely translated version, using fictitious names:

> According to the court decision of Gurlan district on 28 August 2001, Marat Karimov was sentenced to three years of correctional punishment, but after some time he was appointed head of the Fermer and Dehqon Association of Yangibozor District.
>
> If one gets into the habit of doing something, it is difficult to keep him from it. Marat Karimov did not learn a lesson from the past and again committed crime by abusing his position. Particularly, he cheated in the distribution of funds allocated to the Yangibozor District Fermer and Dehqon Association (FDA) by the regional financial administration, on the basis of the law of the Cabinet of Ministries of Uzbekistan, dated 31 December 2001, on 'Macro-economic figures and Government budget of the Republic of Uzbekistan: Schedule for 2002'. Following the saying 'It's your time and use it completely', he began to use the budget of the FDA for his own interest. In other words, instead of equally distributing 5.5 million *so'm* to the fermer enterprises of the district, he took care of himself and his relatives and committed a crime. Without hesitation, he moved nearly half the mentioned sum, 2.9 million so'm, to the account of his own farm 'Erka Yusuf'. Four hundred thousand so'm

---

[152] Elwert (1995: 112) revisited the concept of political entrepreneurs, although his description of them as 'catalysers of collective switching' diverges from the reading of my case study.

were shifted to the account of the farm 'Soliy bobo', 300,000 so'm to 'Omad', 100,000 so'm to 'Jasurbek', 300,000 so'm to 'Shahkuli', and 300,000 so'm to 'Haytim Avaz'. When the turn of the farm 'Ipak yo'li' came, he showed a bit more generosity by making a 'present' of 700,000 so'm.

Never having been caught by anyone before, he continued to defraud the FDA, thinking that all his crimes would go unnoticed, as in the saying 'It had snowed and covered up all traces'. When the financial department of the *rayon* allotted more than 4.8 million so'm to the FDA for the year 2003, he again did everything his own way. By transferring to his bank account and his relatives' accounts a significant part of the funds of the FDA, he caused great damage to the interests of the dehqon and fermer enterprises of the district. But the illegal tricks of Marat Karimov did not stop there. ...

What made Marat Karimov follow this path again, after having already accomplished unworthy deeds in the past and after escaping a penalty as a result of an amnesty? Is it because of his being a slave to his own desire? Is it only irresponsibility? Besides, how could one who had not served his sentence get such a high position? Didn't the people who appointed him know about his past? Where is our society going if the people who have such responsibilities are free to deliberately commit one crime after another? ...

The tricks of the FDA employee Jumaniyoz Omonboyev surpassed the deeds of his master. He was an accomplice of Marat Karimov's in the illegal distribution of the above-mentioned funds to farmers. In other words, together they became involved in a criminal business. On 18 August 2003, Jumaniyoz Omonboyev came to an agreement with chief accountant Rozimboy Jumaniyozov and thus continued the deeds of his 'instructor' during his temporary fulfilment of his duties as an FDA director. Again, out of more than 4.8 million so'm [allocated to the FDA budget by the central administration] ... he managed to transfer 592,000 so'm to the fermer enterprise 'Yadgor khoja', 903,000 so'm to 'Pirnazar ota', 968,000 so'm to 'Soliy bobo', and 405,700 so'm to 'Ipak yo'li' to cover their debts with the input-supplier 'Agroxizmat', caused by additional expenditures related to the cotton-growing and seed production of these farms. In total more than 2.8 million so'm were transferred to these fermer enterprises. ... Neglecting the law and following the saying 'It's your time, use it completely', Marat Karimov and Jumaniyoz Omonboyev acted only for their own interests, and they paid a penalty.

The article was published in a local newspaper addressing a readership of agricultural professionals, and it presented evidence (some of which I have left out) for the accusations. It was signed by the prosecutor of neighbouring Gurlan district, who most likely had directed the investigations. My acquaintance's intention was to warn me about some of the persons he and others in the agricultural department saw me dealing with. Indeed, the two men mentioned in the article were well known to me. I had met them during my first visit to Yangibozor the previous spring, and they were among those who had helped me in the initial phase of my fieldwork by introducing me to fermers on both formal and informal occasions. During the time of my stay, a friendly, confidential relationship had evolved between us. This fact was well known to many in the district, so that at some point I began thinking that the hokim's reluctance to receive me might be linked to this friendship.

I did not manage entirely to clarify the reasons for Marat's conflictual relationship with the hokim. As far as I could reconstruct, after the cotton harvest of 2003 he resigned as rais of the FDA. When I asked the reason for his disagreement with the hokim, he said he had complained to the hokim that his salary was not being paid. He also complained about the hokim's unwillingness to reimburse him for the fuel he had to buy to conduct his daily tasks as FDA manager, an expenditure that greatly exceeded his salary. The requests degenerated into a quarrel. While explaining this to me, he declared in a moment of anger:

> I approve of the politics of our president. But the people below him have totally lost their sense of proportion. I worked as rais of the FDA for three years. I organised the passing of the shirkat land to the fermers, organised bank loans for the fermers, I studied the legislation, and I organised the transition. I did the propaganda [*agitatsiya*] among those who were hesitating, in order to convince them to take the land. And now they put me out in the cold. The reason? It is all about corruption. Those in power want to earn shamelessly. They invest thousands and want back billions.

Another version of the story, given by another FDA employee, contradicts Marat's reconstruction of events. According to it, Marat had to leave after three years of service when his placement was not reconfirmed by the regional (*viloyat*) administration, officially on a formal pretext but in fact because he 'was never in the office and cared only about his own business'. It was 'the hokim alone who organised and implemented the reform in winter 2002–2003'. After that the hokim made Marat the head of the inspection department, one of the district's administrations, where he remained briefly. Two months later the quarrel took place, ending his career in the district administration.

Published several months later, the article added new elements to the story, further complicating these mutually exclusive explanations of events. Evidently, the prosecutor's accusations were aimed at reinforcing the attacks against the former FDA rais, now in disgrace. My initial doubts about the accuracy of the article were due to the lack of press freedom and the pro-regime use of censorship, but later someone in the FDA confirmed most of the facts, although he denied that the farms involved had any kinship relations with the two accused men, as was stated in the article. Moreover, this person raised doubts about the criminality of what he considered a widespread and accepted practice of budget management ('This is usual practice'). For him, the 'guilt' of the accused consisted not in embezzlement but in some (not clearly motivated) disapproval by the authorities, the reasons for which the speaker was ignoring.

It was impossible for me to clarify what interests, actions, and strategies animated this little political scandal in the district. Most of it must be left to conjecture. In light of the arbitrariness that many people attributed to the power holders, some of the article's statements – such as 'Where is our society going if the people who have such responsibilities are free to deliberately commit one crime after another?' – assume an ironic flavour when expressed by the accusing writer of the article. As in many other cases in which I attempted to reconstruct local narratives of conflicts, the facts exposed in the article brought up new questions rather than resolving old ones. If the former district official had a record of criminal conviction, why had he been reinstalled as a rais? If he had been found guilty of a crime, why was he given only mild administrative sanctions? Why was a certain type of deviance (a type of public spending as done by some) emphasised while others (such as the authorities' interference in fermers' affairs and the retention of salaries by the heads of administrative bodies) passed by unobserved? How much resonance did these accusations have among the people of the district and among the agricultural elites?

In the past, conversations with the men accused in the article had contributed to the formation of my views on power dynamics in the district. It was Marat who, in a situation of particular anger, presented current conditions as a degeneration of the political habits of the past: 'If under Sharof Rashidov corruption was 50, today it is 100', and 'At that time anyone found guilty of a crime was sent to Siberia; now they are by themselves [meaning that they could count on impunity]'. These statements, in light of the accusations exposed in the article, took on a somewhat different meaning, yet the parallel Marat drew between the time of the cotton scandal and the present situation seemed to point towards a deeper affinity of causes between 'then' and 'now', causes rooted in the cotton economy. Yet the cotton scandal and the present-day battle for cotton, respectively representing the critical points of the Soviet and post-Soviet cotton economy in terms of relations among communities, local actors in agriculture, and state elites, arose from different circumstances, followed different patterns, and gave rise to different conflict dynamics.

A similarity between then and now is the central role that the agricultural bureaucracy continues to play in many illegal or 'parallel' activities. Today, as in the past, power issues in agriculture develop in and around the realm of agricultural bureaucracy. But today the principle 'It's your turn, so take advantage of it', as illustrated in the news article, is emblematic of the way in which the local political system has, in the perceptions of many people, degenerated. The principle seems to fit everyone, both accused and accusers. In local perceptions the difference seems to be that the weak, the 'losers', are attacked and condemned for what the winners, or strongmen, can do with impunity. Deprived of the moral justification that those charged with corruption could claim for themselves during socialist rule, the accused now earn less sympathy and understanding for their actions than their counterparts did in the past. Behind the conflict-free façade of official representation, local power dynamics in reality develop in the oblique modality of political intrigue. For an outsider, the result is that firm methodological limits exist in attempting to study these political intrigues – a circumstance rooted in the present political order, which does not tolerate antagonism carried out openly in the public sphere.

A difference between past and present is that local bureaucracies operate under greater restraints today than during the 1970s and 1980s. An example is the Yangizobor branch of the Fermer and Dehqon Association. Despite the cliché of pervasive corruption characterising post-independence government organisations, one would expect the FDA, with its mandate to assist fermers and dehqons with advice and legal counsel, to be at least a well-functioning clientelist network able to channel resources to its affiliates. In Yangibozor the reality was more modest. The situation I witnessed concerning one FDA rais, who held the position some time after Marat, was revealing of the changing conditions in the district. At the time, Rajabboy, formerly the land measurer for the shirkat Jayhun, served as the temporary rais of the FDA, having been appointed to the position by the district hokim in an effort to rescue the FDA from its poor state of affairs. Busy with that work, Rajabboy was neglecting his own fermer enterprise in the summer of 2004.

While Rajabboy and I were drinking tea in the office of the FDA one afternoon, a woman came in and interrupted our conversation. She had been employed by the FDA for a couple of months under one of the previous raislar, and now she had come to ask whether she could get something in compensation for the salary she had never received. After listening to her, Rajabboy got a bit angry, saying that many others, too, had received nothing so far. She would receive compensation in due time, when it was her turn. At the time Rajabboy was overseeing the revision of all FDA financial transactions made by his predecessor, in order to protect himself from possible accusations of mismanagement. The woman urged him, saying that even a chicken would be fine, because her family needed the money. Rajabboy insisted that he had just been appointed a few weeks before and had other, more important things to settle first. After the woman left empty-handed, Rajabboy explained to me that in the preceding couple of years, 35 different persons had worked temporarily at the FDA, and 'all of them have requests for salaries'. But it was impossible, he added, 'that people work for some months without any result, basically doing nothing, and then want to be paid'.

The situation was revealing of the mistrust, fear of legal accusations, difficulty of efficient management, and lack of funds that characterised the FDA. In comparison with the administrations of the past, the FDA – standing for many other weak organisations in the district – had exhausted the redistributive capacities that once could turn administrative organisations linked to agriculture into multi-functional power platforms. Whereas one can say, with Verdery (2002 [1991]), that this was a crucial characteristic of 'real existing socialism', the situation today, as exemplified in the cases of Marat and Rajabboy, places new constraints on local political entrepreneurship, which gives rise to a new dynamic.

Today the cotton economy is as much a site of struggle and resistance as it was at the time of the cotton scandal, but there also are significant differences. Unlike the past fraud in cotton, the large scale of which was nourished by the converging complicity of local communities and local and higher-level elites, nowadays fermers, communities, and district authorities are enmeshed in a multitude of little scandals that, more often than not, result from the three groups' diverging interests. Seen from the Uzbek periphery, the socialist cotton scandal resulted in a reflexive forging of a national consciousness that until then had only a 'superficial and fraudulent character' (Carlisle 1991: 119). Today the state authorities use national interests and nationalist rhetoric to justify placing constraints on the individual entrepreneurship that motivates the battle for cotton. In a context in which political scandals multiply and the effectiveness of anti-corruption measures is diluted, as seen in Marat's story, the many local scandals of the agricultural bureaucracy have become a safety valve for power holders, sign of a political leadership with weakened credentials of legitimacy.

Nevertheless, in the context of my fieldwork, the hokim Ergash Yusupov's firm hold over the district restrained the public display of 'rebellion rituals', at least at the level at which 'leaders intrigued against one another for power around the king ... and some princes intrigued for the kingship itself' (Gluckman 1959: 77). Friendships and allegiances were not displayed ostentatiously, and coalition-building was kept concealed, because coalition-making beyond the framework of the legitimate state hierarchy was considered an affront to the authorities.

As in every political system, so in Yangibozor conflicts and struggles are ubiquitous. But there, when they are directed 'upward', they must be kept hidden or indirect – transformed into gossip or stylised in musical performances, for example – or they will simply be nipped in the bud. Khorezmians duly say, 'May the khan not be your opponent in court [da'vogar xon bo'lmag'ay]'. The social ferment expressed in the new patronage relations, however, shows that a lack of open conflict has not prevented social dynamism from taking other forms. Even the primacy of the 'khan' over a district can be challenged by the individualistic activity of incumbent sponsors.

Concluding Remarks

Has the post-Soviet transition in Uzbekistan reinvigorated an informal system of governance based on clan-like patronage networks, as posited by Collins (2004: 224)? Do the patronage networks at work indicate the perpetuation of Soviet or even pre-Soviet patterns, as many scholars have tacitly assumed? Post-reform Yangibozor, where the informal structure of the past steering and control system has opened up in parallel with the reform of the production system, is a good vantage point from which to follow up these questions. Empirical findings seem to minimise the usefulness of 'clans' (Collins 2004; Schatz 2004) and 'solidarity groups' (Roy 1997b) as analytical categories for understanding informal political processes in Yangibozor. Attention to more territorially linked dynamics better captures a too often misinterpreted and misrepresented local dynamism. Radniz's (2005) acknowledgement of the 'power of the local' and Fumagalli's (2007) attention to 'local authority figures' seem to address such processes better and come closer to the form of individualistic entrepreneurship I observed in Khorezm's agricultural world. Yet the novelty emanating from Yangibozor is that the kolkhoz – despite some scholars' and some governments' attempts to reanimate it – has really come to an end.

Sociological and anthropological analyses of socialism and postsocialism often highlight the way the kolkhoz was always more than just a large-scale collective agricultural enterprise (Hann and the Property Relations Group 2003; Humphrey 1983; Petric 2002; Visser 2008). It pervaded the lives of its rural inhabitants as a 'total social institution' that encompassed the whole range of their political, cultural, and economic relations. As an emanation of an ideology and of a state-building project, the kolkhoz acquired the characteristics of a system of governance, able to impose a uniform trajectory upon the most disparate local realities of the socialist sphere. In practical terms the system of governance represented by the kolkhoz was the reality of socialist rule: it socialised the rural citizen through the propagation of an implicit model of power relations, nominally egalitarian but in fact hierarchically stratified.

For many theorists of (post)socialism, the specificity of this system of governance lay in its operating principle. Unlike in capitalist relations, the kolkhoz was governed by a redistributive dynamic, the motor of which was what Verdery (2002 [1991]) called 'bureaucratic allocation', referring to the capacity of individuals to exert control over the distribution of resources by virtue of their affiliation with the apparatus of state bureaucracy. The socialist model of redistribution is not that described by Polanyi (1957), but the resemblances are fairly strong: 'In the redistributive system commonly described by anthropologists, chiefs redistribute goods to their followers, just as socialist bureaucrats allocate social rewards' (Verdery 2002 [1991]: 372). Verdery further observed that 'like chiefs in such redistributive systems, bureaucrats are constantly under pressure not to be outdone by other bureaucrats: they must continue to strive for influence, amass more resources, and raise the standing of their segment of bureaucracy' (2002 [1991]: 372), as if the engendering of a specific dynamic of competition among chiefs and bureaucrats was an intrinsic characteristic of the system she outlined.

In this respect as well, the kolkhoz seems to have displayed a remarkable degree of uniformity across the socialist world. The Central Asian kolkhoz was no exception. The end of collective agriculture and the switch to a progressively more market-oriented economy is also having repercussions in Khorezm on the vacuum that the demise of the kolkhozes left in the local organisation of politics. My research in Khorezm indicates that one can speak of a switch from a straightforward system of patronage during the kolkhoz period to a more open but no less constraining market of political protection in which those who were once the controllers of agricultural resources are still closely involved in the process of production.

Chapter 7
Farmers, Communities, and the State

One of my early field note entries offers a snapshot of the way relations among agricultural producers had evolved by 2003, mentioning the most important players, their stratified relationship, some old and new rules, and discrepancies in wealth and power:

> In the early morning I am already in the car 'on the hunt' for *fermer*s. Although in recent days I have had bad luck, I know today I will not find empty offices, because this is the time of the year when fermers have to deliver the SX-4 form (once a duty of the kolkhoz, now of the fermers), and the district-level headquarters of the agriculture department turns into a stage where everybody wants to attend, meet people, maybe hand out invitations to an upcoming circumcision feast (*sunnat to'y*) to peer fermers from the neighbouring village. I arrive at the former Agroprom building, which now hosts a wide range of 'new-but-old' agricultural organisations, and there is no place to park the car. Everywhere Nexias and Volgas, and old Moskvitches and Zhigulis. The fermers I could never find in their fields and with whom it was impossible to get an appointment, and whose farm workers never knew where they were and when they were coming, here they are!
>
> They are here queuing to negotiate with their boss (*yoshulli*) for the SX-4. This is the period of the year, after having taken the inputs for the state crops from the input supply agencies and having sown the cotton, in which fermers can try to modify their cropping scheme and diminish the cotton quota they had in the sowing plan in January. Such reduction makes the subsidised fertilisers available for the more profitable rice growing of the summer season. But the boss has to give the green light.
>
> When ... [the boss] arrives, he greets the crowd and shakes the hands of the other *raislar,* easily recognisable because they mostly wear green hand-tailored boots, distinctive for officials like kolkhoz

directors and agronomists, for people who have a say in the agriculture of the district, whereas they would look ridiculous on others. ... On the way back, along the road from Yangibozor, in almost every field, rows of peasants, mostly women with their families, are busy with the weeding (*qatqaloq*), muffled-up figures bowing over the small cotton plants in an already warm, late-spring landscape.

This sketch gives a taste of the setting of Khorezmian agriculture after decollectivisation. It describes a moment of interaction between boot-wearing administrators and queuing fermers over the definition of their farms' cropping schemes (*yer balans*), the moment of the year when the fermers' business plans, in accordance with the season's whims, finally become binding directives. The fermers wait patiently to be able to make favourable or at least viable deals for their fields. The delivery of the SX-4 form is also an opportunity for fermers to meet each other, and there is a sense of collegiality among them. Their yoshulli, on the contrary – the head of the district department for agriculture (*boshqarma*), who signs their forms and interacts with fermers about their farm arrangements – greets the crowd but shakes the hands only of other important people, such as the raislar. The farm labourers, especially the women observed working in the fields, are *gektarchi*s and *dehqon*s. They are not part of the meeting, and their absence is telling of how little say they have over agriculture.

Post-independence land reforms and agropolicies in Uzbekistan have affected the organisation of production, the conditions of livelihood in the former kolkhozes, and the forms of interaction between actors in agriculture. Since decollectivisation, the emergence of a new, deeper division between dehqons, fermers, and state authorities has been the most important social phenomenon characterising the rural situation in Uzbekistan.

Having examined the way social relations evolved in Khorezm during the Soviet period (chapters 2 and 3) and discussed the way Uzbekistan's reform policy has affected rural society by reshaping rules and deepening divisions (chapters 4 and 5), I looked at the political struggles over local farming that have evolved in the context of decollectivised agriculture (chapter 6). The questions addressed in this chapter pertain to the moral reasoning about the new statuses, struggles, and dynamics triggered by decollectivisation among dehqons, fermers, and district authorities. Whereas the model of rural society centred on the kolkhoz was maintained largely unaltered by a conservative policy throughout the first decade of the post-Soviet period, during the second post-Soviet decade decollectivisation has been transforming that model more substantially. After its 'anomalous' beginnings, Uzbekistan seems lately to have realigned with the larger scenario of postsocialist rural reform portrayed by Hann and the Property Relations Group (2003), in which decollectivisation has been accompanied by various forms of rural reaction to (or against) increased market pressure and (more) exclusive ownership rights.

Hann asked whether rural communities were fading away under the pressure for stronger economic performance triggered by the 'egoistic' opportunities created by access to markets, or whether rural people, 'despite radical changes in ideology and legal regulations ... actively strive to hold onto the valued elements of the older moral economy' (Hann 2003: 37). In this chapter I attempt to see how far these questions can be answered in terms of the ongoing transformation of Khorezm's rural society, by focussing on fermers', dehqons', and district authorities' diverging attitudes towards and expectations of the agrarian transition. Are rural communities being eroded under the pressure for greater productivity and a more rational allocation of resources? Are fermers exploiting the communities, and does decollectivisation therefore change the 'glue' of the community? What is the role of local administrators in this process? Do state authorities and fermers feel a sense of responsibility towards the communities, and if so, in which direction is it evolving?

For dehqons, recalling the welfare-centred kolkhoz 'spirit' is appealing, but their quests and contestations are unconnected to the idea of reviving the kolkhoz from its ashes. Rather, they raise issues of distributive justice. Fermers, on the other hand, distance themselves from a kolkhoz-style moral economy, which they view as an obstacle to the more market-oriented freedoms to which their new status commits them. For its part, the government makes political order and stability its priorities, rather than distributive justice or market freedoms. Thus it selectively uses either the older moral economy discourse of the kolkhoz or a discourse promoting an essentially state-led modernisation that only superficially resembles free-market capitalism in order to justify its policies. As its implementers, district authorities are called on to successfully handle both discourses.

I begin my discussion by introducing the academic debate over rural community development in Uzbekistan. I then illustrate the changing points of view and conditions of dehqons, fermers, and district authorities and address the evolution of mutual relationships among those actors.

A Broken Social Contract?

Rural communities' relationships with the former kolkhoz and with local authorities have changed since decollectivisation. Deniz Kandiyoti and Olivier Roy represent two influential but opposing views of the dynamics of rural communities' recent transformations in Central Asia. As a starting point, both authors depict the Soviet kolkhoz as a patronage system in which political loyalty, material benefits, and resources were channelled through personal networks, behind a façade of formal rules. In Uzbekistan these personal networks were locally rooted and, as Roy (1997a) in particular stressed, derived a large part of their legitimacy from their proximity to local 'traditional' social institutions. Details of the organisation and modes of operation of these networks have been known at least since perestroika. For example, the investigations of the public prosecutors Gdlyan and Ivanov drew public attention to the previously mentioned 'Uzbek affair' (Gdlyan and Ivanov 1994). Although theft was widespread, local perceptions of the kolkhoz system were generally positive as long as everybody profited from the theft.

This changed after independence. Although with the end of the Soviet Union the 'victims' of the cotton scandal were rehabilitated, the decline of living standards, which especially affected rural people, brought general disaffection with the post-independence reform scenario. According to Kandiyoti (2002, 2003a, 2003b), today's perception of discontinuity with the Soviet past is rooted in one or several 'broken social contracts', in contradiction to what could be called the 'implicit local deals' characterising Soviet times. One such broken contract involves the relationship between the central authority and the regional elites:

> The give and take necessary to ensure the loyalty of regional bosses, by providing them with the wherewithal to extend patronage to their own constituencies, has been one of the driving forces behind domestic policy. The social contract of the Soviet period has progressively been exposed to new sources of strain: the regional bosses and patronage networks were increasingly deprived of their share of cotton export revenue (Kandiyoti 2003b: 145).

On the local level, the other, perhaps more crucial, broken social contract is the one that offered kolkhoz workers adequate formal and informal compensation:

> An understanding of changing entitlements must necessarily take account of the type of social contract represented by the Soviet farming system. In line with the more general principle of 'labor decommodification' ... wages of collective workers were always low but compensated for by a bundle of social benefits channeled through membership of enterprises, including access to a plot for household use. These formal benefits were complemented by more informal mechanisms of paternalistic responsibility vis-à-vis workers, such as helping them to defray the costs of life cycle ceremonies or assisting those stricken by disease or personal tragedy (Kandiyoti 2003b: 158–159).

Kandiyoti painted a picture of winners and losers in the Uzbek agricultural reforms that is loaded with tension and resentment on the losers' side. This situation arose when the paternalistic commitment of the rural elites towards their clienteles and the state policy of laissez faire towards local arrangements in the kolkhozes ceased to exist, upsetting the balance between the exigencies of the local communities and the policies of the central state. These changes first appeared after independence as the glue of rural community life, its links of reciprocity and solidarity, was eroded by the effects of the new reform policies. In this context, 'the combination of high population pressure, loss of nonagricultural employment, changes in cropping patterns and rapid commodification of land have eroded the social contract between farm managers and their workforce and are creating a source of deep discontent' (Kandiyoti 2003b: 154).

Kandiyoti's emphasis on the erosion of social ties in the community contrasts with Roy's account of the perpetuation of traditional solidarity groups through the kolkhoz. Roy's central idea was that a recomposition of traditional forms of solidarity took place within the kolkhoz, pushing aside the original Soviet project of social engineering, which aimed to penetrate and de-traditionalise local society and thereby dissolve the recalcitrant Central Asian rural communities into the wider Soviet society (Roy 1997a). The idea that traditional forms survived and adapted to the kolkhoz was not new, nor was this phenomenon unique to the Central Asian kolkhoz (cf. Humphrey 1983). New, however, was the idea that the kolkhoz should be considered an indigenous, Central Asian form of civil society, a local social fabric able to generate and preserve communities (Roy 1999). In contrast to Kandiyoti's account, for Roy the kolkhoz was, although economically and socially conservative and traditional, the local sphere of participation and a form of lived community able to preserve rural society from the trends of decomposition activated, after the end of the Soviet Union, by markets and privatisation. He asked, 'Is the kolkhoz just a legacy of a centralized and statist system, or did it acquire some social and economic autonomy in the transition period after independence? Is the kolkhoz doomed to disappear in favor of privatization, or does it retain a kind of social personality which makes it an actor not only in the transition but also in the emergence of a civil society?' (Roy 1999: 109).

Roy and Kandiyoti gave different answers to the question of how the end of the kolkhoz had affected rural communities. Whereas Roy argued that the preservation of some form of 'kolkhoz society' was beneficial to local communities, in that it might alleviate the social burdens of change, Kandiyoti believed that traditional social cohesion had already faded away, thus heightening social inequality and undermining the communities.

I believe neither Kandiyoti's nor Roy's interpretation entirely applies to Khorezm, a topic to which I return at the end of this chapter. Meanwhile, in the following sections I turn to the points of view held by dehqons, fermers, and district authorities about the ongoing transformation of rural communities. In the course of decollectivisation in Uzbekistan, those included in post-collective agriculture and those excluded from it – fermers and dehqons, respectively, as described in the land code of 1998 – have been shaped as distinct categories. The creation of these two groups has reaffirmed and deepened a division of labour and roles that already existed in the Soviet-period kolkhozes.

Dehqons' Complaints

The origin of today's dehqons can be traced back to the waged labourers (*kolxozchi*) of the kolkhozes. Although vague and residual, the term *dehqon* is not casual. It was chosen by the Uzbek government for its ideological reference to the 'historical' sedentary Uzbek oasis dweller, who made his living from agriculture. By playing on the notion of the proud, sedentary Uzbek peasant before collectivisation, the term evokes reconciliation with the pre-collective mode of living and with national roots. At the same time, it conceals behind a rhetoric of cultural legitimation the dramatic worsening of conditions of the lower class of former kolxozchi in the years since independence.

Repeated in propaganda posters and in the speeches of government officials, the message is one of ideologically linking the agrarian reforms with a pre-established model of rural society, in the hope of inculcating the idea that the new model, too, is viable. Interviews at the social department of the Yangibozor district administration (*hokimiyat*) and with staff of the Fermer and Dehqon Association (FDA) confirmed this, in that in the eyes of the authorities, *dehqon* did not have to be a synonym for poverty. According to the head of the social department for the region (*viloyat*), an abundant workforce, the pooling of *tomorqa* plots through married sons, the knowledgeable management of agriculture, hard work, and sacrifice enabled most rural families to live economically decent lives, although admittedly poorer than in the past. The emic view of a dehqon in whose house I spent some hours during my survey did not contradict the authorities' description. In response to my question about what he considered to be a good life in his village, he said, 'To eat meat every day; to be able to put a bottle of vodka on the table when a guest comes; to make the *to'y* as custom commands; to build a house for my sons; to live in peace with the village'.

Plate 14. A wall of propaganda. The inscriptions say, from left to right, 'Land is the chest, water is the gold'; 'Cotton is our national riches'; 'If wealth comes from labour, life will be beautiful'; 'A man's greatness comes from his work'; 'Love for the motherland is a matter of belief'.

Since the end of the shirkats, even this modest ideal of dehqon life has become difficult for a growing number of rural families to attain. In discussions with dehqon family heads who at the time were working as sharecroppers (*pudratchi*) for fermers, the idea of the dehqon as an attractive or desirable model was rarely expressed. Everyone among the dehqon households who lacked additional land beyond the tomorqa plot complained that the tomorqa was insufficient to meet the family's needs, and that cash was lacking. Widespread among the former kolxozchi was the belief that conditions would worsen for dehqons and that a future in agriculture was possible only if one became a fermer. Yet in interviews most of these 'losers' of the agrarian reforms ignored the question of their future prospects, finding themselves trapped between the acknowledgement that conditions were worsening and the recognition that alternatives were lacking.

As revealed by my household survey, very few dehqons considered it possible to become fermers in the future. Their reactions to the redefinition of the roles of dehqons and fermers ranged from resigned acceptance to anger, depending on the degree of trust they placed in my promise of confidentiality. Doing fieldwork in the Ferghana Valley in 2004, Zanca (2010: 77) observed that 'the level of dissatisfaction and anger with the state and the leadership coming from rural areas had grown to the extent that people complained more openly than ever before'. My experience in Khorezm the same year confirmed this. Representative of this mood was a conversation I had in August 2004 with a pudratchi, about 50 years old, whom I knew because a colleague had carried out experiments on the fields on which his family formerly worked:

TT: So how is it, to make a living in agriculture [as a pudratchi]?
CF: If they would leave agricultural production to the people [*xalq*], this would be okay, but they don't let them. They should pay us, but they do not. For instance, when they owe us 13,000 *so'm* [per month; US$13 at the time] during the growing season, then they say that they don't have the money. They pay in kind [*mahsulot*] or with the *chek* system [vouchers for non-perishable foodstuffs from state retail shops], but even then the prices are higher than in the bazaar, and so they don't pay anything [they pay less, and in kind]. Only during the cotton picking do they have to pay something; otherwise nobody would work. But if the rule is that they have to pay 36 so'm per kilogram, they pay 28. Therefore everybody leaves for Russia. Thirty per cent of all men are there. Here there are no jobs. A son of mine also went there this year; there he grows watermelons. ...
TT: What do you want from Tashkent?
CF: From Tashkent we want a factory with jobs. Before [during the Soviet period] everybody always used to work. Today the young men are on the street doing nothing. There are dozens of young men, your age, without work here. They drink, take *nos* [a sort of chewing tobacco], they don't know where to go. Before, there was no such thing.

The interview highlighted the way dehqons did not buy into the essentialist ideal of peasant livelihood propagated by the authorities' rhetoric, but rather longed for the security of employment and an orderly prospect for the future.

The prevailing perception among the lower classes of the former kolkhoz was that the state had betrayed and abandoned rural people. 'The state is not present for us anymore [*Hozir davlat yo'q bizda*]', the geography teacher O'ktam once commented, referring to the decay of salaries and services that had especially affected rural workers. As a response to growing hardship, the government had initiated a social welfare programme channelled nation-wide through the neighbourhoods, or *mahalla*s. Village administrations receive a small amount of money that could be given to families facing particular hardship. However, as Kamp (2004) recognised, the post-Soviet social welfare programme was insufficient and exposed to the arbitrary power of the community heads who controlled the resources. In Khorezm, although the programme was perceived to be good in principle, it was only a 'drop in the desert' of people's needs, and it did not change villagers' perception that state responses were insufficient.

The authorities style the dehqon as a viable model, but villagers complain that with the growing hardship of everyday life, the model is not viable at all. The problem with the reforms is that the country has gone through a retrocession, and rural people have been most affected by it. Health, mobility, education, and gas and electricity met high (Soviet) standards in the past but have now become a problem, in the villages even more than in the cities.

In the past, opportunities for upward social mobility through higher education were within reach of people in even the remotest villages. The former kolxozchis complain that this is no longer the case, and education has increasingly become a privilege of the rich. As a university teacher in Urganch who had a village background put it, 'The village offers no alternatives in-between the plough and the pen'. In the past, unlike city pupils, who were more attracted by the opportunity to make money in the bazaar than by university studies, pupils from rural areas were motivated to continue their studies at university. Now, when even the cash necessary for proper clothing to send their children to study in Urganch has become a problem, not to mention the newly introduced university fees and the cost of transportation, dehqons face financial problems while fermers' sons and daughters can more easily afford to go to the city to study.

Although dehqons were excluded from the big land redistribution in 2004, I found no evidence comparable to the double 'no' to decollectivisation and private farms that Hivon (1995) found in rural Russia in the aftermath of the kolkhozes. In Khorezm I found that dehqons generally accepted decollectivisation and did not question it in its own right. Also, after more than two generations, little memory of pre-collectivisation ownership claims by former land-owning elites in Khorezm was preserved. In an interview during my survey, this topic came up with the head of a former land-owning bey family, who worked as an ordinary dehqon and was married to a nurse. He said that 'although we were rich, it is impossible to get back the land. There are no papers anymore, and also the kolkhoz has changed everything'. Nor were demands for alternative forms of land distribution emerging among dehqons in the villages of Yangibozor in challenge to the implementation of decollectivisation. Some dehqons I interviewed recalled with nostalgia the period of the Soviet kolkhoz, but I observed neither the articulation of a counter-programme against the new system nor a wish to reinstall the kolkhoz. Many dehqons, however, expressed their discontent with the fermers, although in low voices and undertones. After I concluded a long interview with another pudratchi, he told me that I was asking the wrong questions. For future interviews, he said, I should add the questions, 'How do the fermers pay for the work? How do the fermers give orders? Was it harder to work in the kolkhoz or is it harder to work for a fermer?'

These were rhetorical questions alluding to the fact that, contrary to the situation on the collective farm, fermers often did not pay wages on time or in full, were not as knowledgeable about agriculture as many of their employed workers and consequently often gave incorrect work instructions, and squeezed the dehqon workforce because they wanted to maximize their profits. 'If you don't do your job well, the fermer will fire you', the man added. Although dehqons expressed resentment and envy against fermers, whom they accused of caring only about their own profit, strikingly, no idea of 'resistance from below' had emerged. As far as I could judge, Khorezm's 'weapons of the weak' (Scott 1985) seemed to be fairly blunt. When asked what he would do about a situation he described as bearing no future, this dehqon laconically answered with a variation of the previously mentioned khan metaphor: 'If the judge is beating your mother, whom will you tell about it [*Qozi opangni ursa, kimga aytasan*]?' Claiming justice against the judge is not within the claimant's reach.

In occasional encounters in which interlocutors felt more protected by anonymity, complaints about the situation of the overwhelming majority of rural people were even more explicit. A pensioner of the former kolkhoz Madaniyat said, 'All the talk is about the fermers, but what about us? How can I feed my family? We are 15 people at home; for bread alone we need 3,000 so'm a day. This is 90,000 a month. Where should I get it from? We can't earn it, we don't have it'. Another peasant, in his forties, who heard about me came to look for me in the fields where I went regularly with a project colleague. He wanted to tell me, 'All is getting bad here. Things are more and more expensive; we do not make it. Look, our village is empty. All the men have left for Russia. Even rice is being grown by our women'. After saying this, he refused to reveal his name, a sign of the fear in which people made such statements.

People excluded from the land redistribution were aware of having been dispossessed of the former collective assets and perceived the land appropriation as illegitimate. A statement I heard several times during interviews in the homes of dehqons and small fermers in Khorezm was, 'The ones who ruled the kolkhoz before, the *inginer*s, *raislar,* and *brigadir*s, they took all the land for themselves and for their relatives'. To what extent this statement reflects the way things really went is another story, but it fits a widely circulating postsocialist rural stereotype according to which post-collective managers 'spout self-aggrandizing accounts; villagers counter with charges of corruption and theft' (Lampland 2002: 46). It reflects the dehqons' perception of the irrevocability of the land distribution and of the injustice resulting from the manipulations that took place during privatisation. Although the excluded do not question the privatisation as such, a common narrative among ordinary rural people is that the appropriation of land by the former kolkhoz notables through privatisation was illegitimate. This goes hand in hand with the perception of the excluded that they have no available options with which to counter a reform process that is unfavourable to them.

Fermers' Ambitions

Whereas *dehqon* is a term from the past, *fermer* suggests a bright future of agricultural modernisation. Semantically, the term seems to be an adaptation of the *fermery* of the Russian reform context (Gambold Miller and Heady 2003). For Khorezmians, however, the notion evokes the Western, American model of farming, suggesting the existence of a real land reform and a real privatisation. It connotes modern, well-equipped, large-scale industrial farming, a connotation that transcends the current straitened circumstances of most farms and looks ambitiously beyond it.

Although reminiscent of Western terminology, the fermer enterprise is modelled more on the kolkhoz and so far should be understood as such. Half a year after decollectivisation, the governor (*hokim*) of Yangibozor district used this metaphor in a speech during the sixtieth birthday celebration (*yubiley*) of a former rais of Bog'olon, Yangibozor's showpiece kolkhoz. To an audience of hundreds of fermers eating *palov* and listening to a performance by the Khorezmian musician Artik Otajanov and his ensemble, the hokim explained that being a fermer was like being the rais of a little kolkhoz, suggesting that, in contrast with the bad habits of the past, with Uzbekistan's independence, 'everybody can be a rais'.

Despite the hokim's politically motivated up-styling of the image of fermers in his speech, most fermers today still resemble, more modestly, the post-reform equivalent of the kolkhoz brigadir, rather than the rais. The brigadir was in charge of the supervision, coordination, and execution of all crucial agricultural operations in his brigade's territory – sowing, weeding, fertilising, irrigating, applying pesticides, harvesting, and so forth – and was a person of authority for the brigade workers and the community. Since decollectivisation, it is the fermer who has taken over this role and is responsible for cultivation, holding a position in the agricultural hierarchy intermediate between that of rais and dehqon.

The much emphasised 'independence' of fermers relates to the fact that every farm has to produce, deliver, and handle its accounting autonomously. In practice, fermers still cannot choose their crops freely but must follow the directives of the district authorities. In this respect, too, the new local producers resemble the old ones. Of course important changes have also taken place: fermers have greater risks, personal liability, and opportunity for economic action. But the fact that most fermers worked previously in the kolkhozes and later the shirkats as executive staff, or are somehow related to the executive staff of today's *motor traktor parki*s

(MTPs), illustrates how the implementation of the reforms entails a resurfacing of already established relations and strengthens pre-existing forms of role distribution.

Whether fermers as a whole could be considered winners or losers in the reforms was not an easy question to answer in 2004, in the aftermath of decollectivisation, especially because of the heterogeneity of the farms subsumed under this one legally defined category. As I showed in chapter 5, after decollectivisation the disparity between fermers was considerable. Only a minority of fermers had large land estates, and farms as large as 100 hectares or more were exceptional. I could count one or two in every former shirkat, and often they belonged to a rais, a former rais, or someone related to a high state official. Large fermers were agricultural notables. Some held high-ranking positions as public officers in the hokimiyat or in an MTP; some had held leading positions in shirkats – for example, as an agronomist or brigadir – and thus had technical university degrees obtained in nearby Urganch or in Tashkent. Others were doctors, militia officers, private businessmen, or, more rarely, university teachers or school directors. Everyone whom I asked agreed that these large fermers (*katta fermers*) were a category apart, hardly comparable to the mass of ordinary fermers in economic status, assets, and, often, social, professional, and educational background. They were either related to the *nomenclatura* of the agricultural production hierarchy or they were 'new farmers' (*yangi fermers*), urban newcomers who had entered the agricultural scene using capital gained in the cities. A decisive asset of big fermers was that they could use their connections, or 'bureaucratic capital', to get good terms of usufruct in their leasing agreements, and often they could mobilise additional resources for their farms through their positions outside of agriculture.

Addressing the social conditions that facilitated the acquisition of property by former collective farm managers during Hungary's decollectivisation, Lampland (2002) applied Bourdieu's social capital argument (Bourdieu 1985). For Lampland (2002: 41), the farm managers' reliance on networks established during the socialist period was not in itself 'sufficient to ensure success in the ever-shifting agrarian market', yet these networks, together with the former kolkhoz managers' expert knowledge and extensive experience in managing collective farms, conferred on them decisive advantages over other incumbents. Although in Uzbekistan, too, the former kolkhoz elites have prevailed over other actors, Lampland's social capital argument cannot be directly adopted, because it presupposes a market environment and a sort of accountability of the farm chairperson towards the shareholders in the former kolkhozes that is absent in Uzbekistan. On the contrary, in Uzbekistan large fermers, rather than relying primarily on knowledge of agriculture, rely on more variegated skills and on social connections in order to encroach on the difficult territory of agricultural bureaucracy. Therefore, I prefer to speak of 'bureaucratic capital', meaning the capacity to influence the process of bureaucratic allocation (Verdery 2002 [1991]).

According to Uzbek law, it is not possible to be head of a farm and simultaneously an employee in a state administrative organisation (*tashkilot*). This apparent obstacle is in fact of little hindrance, because on paper the wives (or sometimes other kin) are the legal heads of the farms while their husbands or other male kin conduct the affairs of the enterprise in addition to their official employment. The district's register of farms lists a considerable number of farm heads as women, but a closer look shows that in most cases this is an indicator of a family with a diversified income structure, in which the head of the household is employed as a public officer but also controls a farm. The side effect of this practice is that the figures for female engagement in agriculture at the management level are, on paper, very high. In reality, farm management has remained a thoroughly male domain.[153]

Even more important than the size of a farm, its 'terms of trade' – its overall conditions and the cropping agreements made with the district authorities – are key to its profitability. Environment, land quality, and taxes play a subordinated role. Good arrangements make the difference between the minority of highly profitable fermers rich in bureaucratic capital and the rest of the agricultural producers. Although national legislation has strengthened fermers' autonomy to choose which crops to grow in which areas and even whether to leave their farmland uncultivated, so far this has showed little effect on the ground. The wife of a fermer summarised the problems of the family farm by saying, 'The laws are good, but our leaders [*rahbarlarimiz*] do not follow them'. This was a sentence I heard several times during my fieldwork.

A high-ranking district officer of the agricultural department justified the constraints on fermers with the statement, 'Our people are not ready for these reforms', meaning that if constraints were lifted, there would be anarchy, and 'fermers would not know what to grow, how to grow it, or where and how to sell it'. The statement illustrates the still pervasive paternalistic approach of state institutions towards their subjects. It was addressed to-

---

[153] In Yangibozor I heard of only one case of a woman running a medium-size farm as a de facto farm manager (*haqiqiy rahbar*), as opposed to a manager on paper, attending the meetings called by the hokimiyat and doing the paperwork at the agricultural department herself. Women's presence and role in agriculture were crucial in the everyday work in the fields, but they rarely held decision-making positions.

wards actors who entered agriculture motivated only by the 'egoistic' idea of making easy money. The way in which decollectivisation was planned and put into practice by the district hokimiyat, as well as the way in which fermers were constrained by the authorities, was aimed at curbing such egoistic trends and keeping decollectivisation 'socially viable' in the terms defined by the hokimiyat.

This attitude of the state authorities puts them at variance with most other postsocialist rural contexts. After the demise of the Soviet Union, in most postsocialist countries rural people's sense of displacement was augmented by the state's retreat from agriculture. In Azerbaijan, for instance, family networks stepped in to fill the institutional vacuum left by the dismissal of the once state-led agricultural facilities, infrastructure, and organisations (Kaneff and Yalçın-Heckmann 2003). In neighbouring Kyrgyzstan, where cotton-growing farmers have enjoyed more freedoms thanks to a more thoroughly liberal reform, rural producers nevertheless end up feeling 'abandoned by the state', and as a consequence, 'cotton growing in Kyrgyzstan is not profitable today' (Kim 2007: 125).

The situation appears to be more complex in Uzbekistan. The dehqons' sense of abandonment by the state derives from their disillusionment about their chances of regaining past living standards and from their acknowledgement that the place foreseen for them by the authorities in rural society is very uncomfortable. Fermers do not feel abandoned by the state, but they feel bullied by intrusive state authorities and view those authorities with suspicion. One fermer said, 'There is never a benefit for us when the state interferes [*davlat hech qachon foyda berib kirmaydi*]'. This is precisely the controversy between the authorities and fermers, as illustrated by the following entry from my field notes, from May 2003:

> We go to Y.R.'s farm. B. from the FDA wants us to go there because he is among the biggest farmers in this former shirkat, growing 35 hectares of state crops. We are invited for *choy*. B. is on good terms with Y.R. and this helps a lot. Y.R. has seven sons, six of them already married. Before becoming a fermer, he was the chief engineer of the shirkat. ...
>
> To my question, what is the main problem in agriculture, he answers: *burokratsiya,* meaning that the hokim exerts pressure (*tazyiq*) to grow the state-ordered crops. 'He is the only one here who can grow rice. ... He gives to himself the rice quota but imposes the state order on cotton and wheat on the independent farmers. Despite of the law of the Oily Majilis (parliament), he bans the growing of rice'. So Y.R. looked up the land code and started arguing against the hokim that according to President Karimov, the state crops can-

not be imposed on the fermers. The prosecutor's office ... intervened when it became clear that Y.R. was serious in his intention not to grow state crops, as there were delays in sowing cotton in his fields. There was arguing at the *prokuratura* and a quarrel between him and the prosecutor. ...

Y.R. had good connections at the prokuratura and could afford such disputes. However, he had to give up and go for the state crops, or he would have had serious troubles. Now he grows 30 hectares of state crops (meaning that on the rest he can grow what he wants). I ask whether he makes profits with his farm and he says he will, but then starts counting all the expenditures he has to face. ... Concerning his land Y.R. says that he got twice as much land as he applied for: not all land is fertile. For him it is inconvenient, because he is forced to grow state crops on unfertile land, with less profit and much more work. Twenty-five former pudratchi families are now employed on his land. Y.R. would have preferred to have less land and fewer pudratchis to look after.

It is in the face of such issues that big fermers attempt to use what I have called bureaucratic capital. When they do not succeed, they must abide by the district authorities' desiderata, which, more often than not, are inimical to their profit-seeking.

The District Authorities' Vision

Talks with the implementers of the government's rural development plans confirmed that small fermers' fears of losing their land were not mistaken. In May 2004, after endless postponements and refusals, I had an opportunity to talk with the Yangibozor district hokim when I met him by chance during his daily rounds in the fields, where he was monitoring fermers, agricultural activities, and the performances of his staff. We talked about the government's desiderata for reform implementation and his personal views concerning the creation of new inequalities in rural society. From the conversation it emerged that in the hokim's plan, farm units were to have been much larger, but the shortage of capable applicants forced the authorities to create smaller farms. I wrote in my field notes:

I ask: 'What will be better, the development of the MTP or of the fermer? Does the MTP have a future or will it become smaller?' Ergash aka smiles and says that of course the development of the MTP is his priority; in future it will become a developed tractor service company. To my question whether the number of farms will increase or decrease in the future and what would be a good size for a farm, the hokim answers in a contradictory way that yes, the number of farms will increase, but '150–180 hectares of land could be an adequate size for a farm'. Then he points to a *kontor* [plot] where he had met with the rais of the MTP and with other local staff before I came: 'Look here, in this context agriculture [*dehqonchilik*] will not develop at all. This fermer has only nine hectares of land, and no means to improve it. Therefore his farm is very bad. Only large farms can make the best of the land'.

I ask: 'How will privatisation proceed?' – 'Completely. Already the stores and the barbershops have been privatised. In future land will be also. Last year we had a new law that the land of the farmers will be inheritable. Now we are leasing the land, but in future land will be privately owned'. ... I ask him: 'What will the others do? What about the rural people who did not start fermer enterprises?' His reply was: 'We have a plan until 2010 that only 35 per cent of the population will work in agriculture'. – 'What about the others?' I ask. The hokim points at his shirt, then at his boots, and then explains in a paternalistic tone: 'Look at this, look at that! Look at bread, milk ... Nowadays the rural families are autonomous in every respect. They do everything by themselves, within their family. How is it in your country? There are people who cook, there are maids in the houses, who do the cleaning and washing, and so on. Therefore I say, those who deal with cotton should deal only with cotton, others should make bread, others should clean, others should become tailors, like it is in your country'.

The future described by the hokim, who seemed to thoroughly represent the Uzbek government's reform plans, held no place for peasants in rural society. Rather, modernisation would entail jobs in industry and services. In this interview in May 2004, the hokim anticipated the farm consolidation policy of 2008–2009, but the accomplishment of the promise of jobs for rural people remained to be seen. With no job prospects except for migration, and under a policy that favoured the development of a few large enterprises instead of many small ones, the weakest elements in rural society were condemned to pay an unequally high price for the achievement of these prospective results.

In the following conversation with an educated dehqon and teacher, representative of the generation most affected by the rural decay, it was, in the 'vocabulary of dissent' of the rural weak (Tingay 2004), the current leadership that had lost most in prestige in the community, relative to the raislar of the past:

A: The raislar of today are not even worth the shadow of past ones. These are persons who do not defend their people, but they are nice to the hokim and say to him, 'Boss, I will manage this [*bajaraman yoshulli*]'. But in fact they cannot! Look at our new rais: for two reasons he is bad. First, he has no agricultural knowledge, he is a house builder–architect [*quruvchi*]. Second, he has not the strength to say no, to withstand and oppose the hokim.

TT: If not he, then who in your village can defend the people?

A.: Nobody! Our rais is weak. In order not to displease the hokim, he will fulfil the plan. For instance, we have to deliver 1,000 tons of cotton for our kolkhoz. They [the MTP staff] will arrange it so that the fermers have to grow for the state plan on 70 per cent of their land, and then the 30 per cent for themselves will not be sufficient to live on. Also, their production is low. But the first priority has to be the plan. Because of one man [the rais] who is incapable, the fermers are incurring debts from year to year. Two years ago the debt reached to the knees, last year it reached the hips, and now it is up to the mouth and the fermers risk drowning because of the debts. The fermers are getting their inputs through bank loans; they pay an interest rate on it. You get credit to produce 15, but then you produce 5, but you'll pay your interest for 15. Look in the channel, there is water. But it never reaches our fields here. A rais should struggle to get it! He should be able to get it for his people!

The teacher gave vent to his discontent with the entirety of the rais's work, not only with regard to dehqons but also on behalf of the local rural economy. In his words, a rais had a moral duty to struggle for the betterment of agriculture and of people's livelihoods. In Tingay's work (2004), the vocabulary of dissent of the rural poor emerges as a tool against landlords' usurpations in the struggle for land in Egypt. In comparison with the situation Tingay described, in Khorezm the voices of the marginalised have too little influence on the power holders of the district. For those excluded from the land redistribution, the insufficient policy response to their needs strengthens their perception of betrayal and abandonment by the state.

This perception of betrayal among the dehqons is linked to the changing attitude of a local leadership keener to serve the command hierarchy than its communities. With the kolkhoz, private ownership of land and means of production was substituted for by entitlements to social benefits and services. Today, for those who have gone uncompensated for the loss of past benefits and services by new entitlements to land or other assets, the curtailing of the past material welfare standards comes close to a new form of fundamental expropriation. Today's generation of excluded persons perceives decollectivisation much the way their grandparents perceived forced collectivisation: as an expropriation of their future. It restricts their chances of leading materially decent rural lives without, so far, offering viable alternatives.

*Fermer*s, State, and Community

Clearly, the burdens and opportunities of the new farming system in Uzbekistan have been unequally distributed. Institutionally, the authors of the reform policy have attempted to maintain intact the social ranking inherited from the kolkhozes, an attempt that has been mediated by district-level authorities as implementers of the policy. As a consequence, rural community change in Yangibozor is not, so far, a story of social desegregation. Even while creating a radical rupture with the past community commitment of the kolkhoz, the reformers propose a re-regulation of the relations of agricultural production in continuity and in harmony with the patterns of obedience, loyalty, and dependence between authorities and producers that characterised the social setting before the reforms. Yet for district authorities, the question of responsibility towards the community has remained an important issue.

One example involved E.A., an elderly pensioner and formerly a distinguished official of the command hierarchy, who became a fermer as a reward for his long engagement at his kolkhoz. After working in the kolkhoz and doing military service in Estonia, he studied accounting in Bukhara and then returned to Yangibozor to work for more than 20 years in the accounting office of the sovkhoz of his home village. In the mid-1990s he became a fermer. 'At that time', he said, 'my eyes were bad. I asked to retire, and my yoshulli [the rais of the shirkat where he worked] said that I could go and get some land from the kolkhoz orchard and become a fermer'. This was a way to ensure him a secure retirement at a time when pensions had lost their value.

The example illustrates the way the reforms were implemented neither to disrupt the patronage linkages that thrived in the kolkhoz system (and made the collective inefficient) nor to select fermers with the goal of enhancing the profitability of land use. On contrary, it shows how links of reciprocal obligation between actors in agriculture at the local level have determined the outcome of land reallocation. The sense of responsibility felt by district policymakers towards their affiliates or subjects may be judged paternalistic, socially conservative, and to illegitimately prioritise those who were affiliated with the kolkhozes, but it contradicts the assumption that the reform is 'predatory' towards communities, aiming only to enhance state revenues from agriculture. If decollectivisation was, from the perspective of the centre, first and foremost a matter of solving problems of efficiency in agricultural production, then from the perspective of the periphery – the district and the kolkhoz – it was a matter of maintaining intact the old social ranking.

It is not surprising, then, that along with the social stratification of the past, interactions between the 'bosses' and the fermers resemble past interactions between district and kolkhoz leaders and the lower-ranking officials. My earlier account of fermers queuing at the former Agroprom building in Yangibozor to do their duty by delivering their SX-4 forms is one example of both continuity and change in the modalities of interaction in the command hierarchy. Another example is the *pokaz,* the compulsory seasonal seminars and meetings called by the hokim or by the rais of an MTP, in which fermers gather to receive instructions on what to do with their fields and when and how to do it. Although the pedagogical utility of these meetings is small (fermers usually know what to do already), they are occasions on which to put the authorities on stage and publicly reaffirm unity and control over the hundreds of fermers who sometimes gather. With the pokaz, the district elites who preside over them establish the rhythm of the agricultural cycle, whereby each fermer's tasks and duties become facets of a greater task that unites the district. Holding these seminars became a ritual in the Soviet Union, and they have been absorbed into post-Soviet agriculture as a legacy of the kolkhozes. Usually on these occasions, as in the past, the hokim blames the raislar for the MTPs' shortcomings, and the raislar blame their subjects, who do not dare counter their shouting, although they were asked to do the impossible. Instead, as fermers told me, not without some irony, they profess their ritual 'Boss, I will manage/fix this [*bajaraman yoshulli*]', meaning that they will preserve the form and face but then do something altogether different, within the realm of the locally achievable.

Although fermers have a privileged relationship with the authorities, their privileges are limited and can be abrogated when the authorities want the fermers to be responsible for their communities. Since the dismantling of the shirkats in Yangibozor, some fermers in every MTP have been forced to return part of their land because a need for new tomorqa plots had emerged in the rapidly growing settlements. Fermers who had leased plots close to the settlements had to return land in amounts requested by the land measurer and the rais. For fermers who had invested social and economic resources in obtaining and enhancing the value of the land, this was a great loss. I asked the district land measurer whether these fermers were compensated in cash or from the MTP's land reservoir, and he replied that they were not compensated at all. All land, he said, belonged to the state, and these fermers had been fortunate to profit from the land for several seasons. In this case, the right of households to their tomorqa plots was prioritised over the fermers' claims to land, which shows how the local government was trying to keep a grip on relations between fermers and dehqons and to keep the sources of tension in the community under control. In many respects, however, this attempt was only half successful, and conditions in the villages have worsened for the poorest part of the population.

The attitude of the authorities towards the massive seasonal labour migration from Yangizobor is telling in this respect. In a conversation with the village militia officer who was instructed to monitor my research activity, I asked why it was so taboo to talk about labour migration with a foreigner like me. He answered, 'It is shameful for our president that villagers have to leave for migration', for this meant that the president was unable 'to feed his people'.

Analogous to the kolkhozes' paternalistic provision of services to rural communities in the past, fermers today are compelled to provide 'voluntary' services to their communities as a sort of collective corvée. In Yangibozor they have had to fund the renovation of school buildings and the enlargement of the district hospital, for which the local government lacked the money. Every fermer, according to his capacity and his economic status, is compelled to maintain a part of a school in the village where his farm is located, according to lists compiled by the chairman of the village administration. It could be a single classroom or the roof, for example. In a village I worked in, the fermers of an MTP were pooled in groups, each of which was assigned a school building that it had to renovate at its own expense. Their work was referred to as *hashar,* a voluntary effort for the community good, or as a donation (*xayriya*).

A 2004 document from the Bog'olon village administration listed the compulsory services that fermers in the four *elatkom*s under the administration's jurisdiction were obliged to perform. State buildings to be renovated – schools, kindergartens, and sanitation facilities – and the fermers of the MTP assigned to the work were listed in adjacent columns. Every fermer was obligated, according to his capacity and in a way determined by the district authorities, to provide for the renewal of a given part of a communal building or to give the equivalent in money. The document was signed by the rais of the Bog'olon MTP and the chair of the village administration.

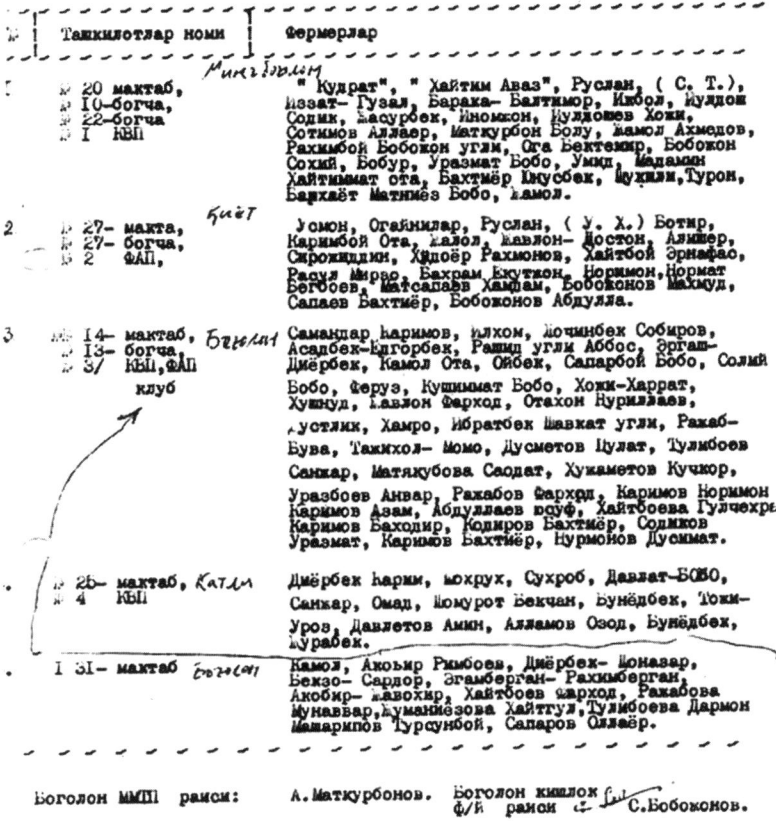

Plate 15. Village council document assigning community corvée to *fermer*s.

Furthermore, a percentage of the fermers' winter wheat harvest was collected and channelled through the staff of the hokimiyat to an ad hoc foundation to finance the rebuilding of the district hospital. Public officers had to contribute to the project as well, and the hokimiyat subtracted part of their salary towards that end. Fermers obey such demands because they know that opposing the hokimiyat will mean trouble for their businesses, although these 'mandatory-voluntary' contributions have no legal basis, and fermers, because of their debts, would not pay if not forced to do so. Some fermers raised doubts about the transparency of these charges and services (some said they were a 'racket' in which the benefits ended up in the wrong pockets), but the practice shows how fermers are identified as capable, wealth-producing actors who should reciprocate with service to their community. The communities see this service as an obvious duty on the part of the fermers. The wife of the geography teacher O'ktam said, 'They are fermers! Of course they should do so!' – meaning, 'If not they, then who else could do these things here?'

In the same fashion the fermers are asked, for public holidays and celebrations, to offer food (*palov*) to the community in a way appropriate to their status and economic means. Palov is a highly symbolic dish because it is usually prepared for *to'y*s and other special occasions, is relatively expensive, and is always consumed communally. During the harvest season, palov figures in the celebration held by each production unit that meets its cotton target. If the target is not met, no palov is offered. Competition arises at the different levels of agriculture to match the target before the others do. Every cotton-growing farmer, every kolkhoz-shirkat-MTP, every district (*rayon*), and every region (*viloyat*) enters into the competition to be the first to meet the target.

Because 2004 had a cold spring, the Khorezm region came in only fourth at the national level, first place going to the Surxondaryo region. In Yangibozor district the first MTP to reach the target was Bo'zqal'a, and a celebration was organised at noon. The fermers sat at two long tables eating palov while a small local music ensemble played in the background. The hokim made a congratulatory speech in which he exhorted the fermers to offer palov that evening to the university students, schoolchildren, and conscripted workers who had helped with the cotton picking. He announced that great celebrations would be held in the stadium once the plan was fulfilled district-wide.[154] Nexias and Damas, Uzbekistan's 'national cars', produced in the Daewoo car factory near Andijon, would be awarded to the best-performing fermers. At Bo'zqal'a, along with the joint accomplishment of the annual target and the common consumption of the meal, 'community' was displayed, celebrated, and symbolically strengthened at the several levels of society.

---

[154] The 'cotton feast' (*paxta bayrami*) was a festivity introduced during the years of the Soviet Union (cf. Chylinski 1991: 160) that is still popular today.

Plate 16. Celebration of the fulfilment of the cotton quota at the Bo'zqal'a MTP, October 2004. The *hokim*, standing on the right, is seen from behind, wearing black. The two uniformed men on his left were the district head of the militia and an officer from the Uzbek intelligence service.

Rural Society after the Kolkhoz

In Yangibozor, relationships between dehqons, fermers, and district authorities were characterised by many contradictions. Fermers emerged as new key actors in agriculture, materially taking over the duties of production and the main burden of risk. Oscillating between the interests of their own enterprises, demands from the authorities, and demands from those who were excluded from land, fermers had complex relationships with their communities. They held relatively high status in rural society, enjoyed the privilege of land usufruct, and exercised authority over their workers, but in turn they were enjoined to perform certain duties towards the nation (growing cotton for the state; 'feeding the nation') and the community (renovation of public buildings; making compulsory donations). In Yangibozor, a large farmer (*katta fermer*) might attempt to show off his status by putting on a *katta to'y*', a fermer's celebration, with pomp and spectacle. Such conspicuous ritual consumption (cf. Koroteyeva and Makarova 1998a) distinguished large farmers from ordinary villagers, whose ordinary *to'y*s were sometimes said to be the only occasions on which the rural poor 'got to eat meat'.

Dehqons, in contrast, although they represent the majority of the rural population, appear to be the residual category in the reforms and have collectively paid a high price, in both economic and social terms, for Uzbekistan's restructuring. Meanwhile the district authorities, affiliated with the restructured command hierarchy, have tried to steer the restructuring and the agricultural cycle paternalistically. Despite their efforts, their capacity to keep social relations on predetermined tracks is becoming increasingly difficult.

Since decollectivisation, a model of community based on a tripartite social stratification – rural notables and authorities, producers, and rural people (the workforce) – has been reproduced and strengthened. Although this model worked well in the past, since decollectivisation it has become endangered by its increasing contradictions. These contradictions reflect the ambiguous mandate of a reform policy that on the one hand aims at decentralising agricultural production and enhancing productivity but on the other hand wants to contain any social process that might bring innovation to a structure of rural society inherited from the kolkhoz.

A state that was once an all-round caretaker has now changed its role vis-à-vis the rural population. In the past the centre's discourse against the local leadership portrayed the raislar as corrupt, but they were accepted locally if, despite their corruption, they were able to redistribute resources. Today's local leadership has lost the unproblematic support of the community. Rural communities previously distinguished between legitimate and illegitimate corruption. They saw the first as necessary and beneficial for the people (as when the kolkhoz's fertiliser ended up on the village's tomorqa plots) and the second as antisocial and directed 'against the people' (as when the former kolkhoz elites appropriated the land and assets of the shirkats). Now that corruption is perceived as increasingly individualistic and entrepreneurial (see chapter 6), its benefits no longer reaching the villagers, the past distinction between legitimate and illegitimate corruption has lost its power to keep relations between people and authorities in balance in rural communities. Rural elites try to maintain the past rhetoric of responsibility for the community's well-being, but constrained living conditions unveil the emptiness of their promises.

Local officials' frequent abuses of their positions and systematic infringement of the law endanger the state's legitimacy and authority in rural society. The impunity of these officials poses a dilemma that only fundamental political reform could solve. Yet much like the Chinese Communist Party in rural China (Zhao 2009), the Uzbek state is unable to find a remedy. Instead, it reacts against this impasse with frequent changes of local cadres (Ilkhamov 2004; Wegerich 2006), who are either publicly accused of or discharged for incompetence, theft, or abuse of office whenever their superiors' confidence in them falls short or local *raison d'état* demands it.

Regarding rural community change, neither Kandiyoti's metaphor of a broken social contract nor Roy's notion of the preservation of the kolkhoz as a solidarity group entirely fits the social processes accompanying decollectivisation that I observed in Khorezm. The social and economic relations I documented in Yangibozor raise doubts about any assertion that local solidarity was produced by the survival of traditional social institutions from the kolkhoz period as the glue of the community. It seems, rather, that constraints imposed on local communities have led external observers mistakenly to the conclusion that the kolkhoz of the past has become a place of social intimacy for communities. Occasionally, group identity rooted in the kolkhoz is indeed put on stage, as during the 2004 harvest celebration I described for Bo'zqal'a MTP, where the postsocialist 'bricks' of the reshaped agrarian hierarchy were publicly displayed in a competition for prizes. Yet in Yangibozor the former kolkhoz better resembles a cage from which fermers and dehqons alike would like to be freed than a place of local solidarity.

Even if the remnants of the kolkhozes do not constitute solidarity groups, this does not in itself validate the idea of the broken social contract. Against that idea, one might argue that the logic of reciprocal obligations and compensations that characterised relations between the different social forces during the kolkhoz period has not totally disappeared. Rather than missing a social contract with the postsocialist state, people remember the late Soviet kolkhoz period with nostalgia as a time when, after more exploitative and socially disruptive beginnings, the subsidies and lax regime of the planned economy permitted local communities to live in some abundance under the Soviet regime. Now that national economic policies and the pace of reform play against them, dehqons' and fermers' fears and suspicions of the state have again grown. That the wealth of the rural communities originated to some degree in the abuse of the economic system by the established native elites (as in the cotton scandal) is another story. The communities' losers under the kolkhoz system, who were excluded from the illicitly acquired wealth, complained about the theft and corruption, blaming local power holders as much as they do today – although being an economic loser today is incomparably worse than in the years of the cotton scandal. Yet against Kandiyoti I would argue that, more than missing the kolkhoz with its social services and benefits (Kandiyoti 2004), people miss the cash or, more to the point, the opportunity for upward mobility through their own efforts.

Chapter 8
Conclusion

With the majority of rural families in Khorezm cut off from formerly collectively owned land, and with a minority of former socialist rural cadres and urban newcomers gaining the most from agriculture, decollectivisation has aggravated rather than resolved underlying social conflicts in this part of Uzbekistan. By exacerbating pre-existing contradictions, decollectivisation is creating new forms of social inequality and exclusion in the Uzbek countryside that fuel an unprecedented rural discontent. Nevertheless, resistance does not take so much place among people excluded from the land reform, where they might most legitimately have been expected, as among those on whom the reform outcome has conferred a more ambiguous place in decollectivised agriculture.

Two distinct sorts of struggles characterise Khorezmian decollectivisation. First, a new social conflict has arisen between dehqons and fermers around disputed issues that affect their livelihood and subsistence strategies and pertain to working conditions, payment, grazing rights, inputs, and, more generally, the devolution of entrepreneurial risks on farm employees and the social responsibility of fermers towards their unrelated employees. Second, a conflict has formed between fermers and district authorities that has no direct antecedent in collective agriculture. It revolves around factors that determine the profitability of agriculture, largely the legal-normative conditions of leaseholds, access to agricultural inputs and markets, entrepreneurial freedom, and the possibilities for farmers to resist the strict control of the district authorities. Although this conflict dynamic can sometimes be explained by the coping strategies and survival needs of the newly created and still economically insecure fermer enterprises, often it hides a struggle over an opportunistic entrepreneurial activity that power holders perceive to be dangerous for the social order and for the structure of power relations in the district.

The difficulties I faced during my fieldwork show that there is no straightforward way to assess the ongoing social and political transformations of Khorezmian society. Nevertheless, in an attempt to disentangle the strings of relationship among the state, rural communities, and local elites, I followed up transformations in regulation, power relations, and relations of agricultural production from collective to decollectivised agriculture. As discussed in chapter 3, the crisis of socialist modernity was rooted in the gap that opened between the social and the material developments of Soviet modernisation: institutions of pre-socialist Khorezmian society – its 'social infrastructure' – attempted to keep up with the growth of society's 'material infrastructure'. The latter was planned, funded, and steered from outside the communities by the Soviet authorities but implemented from within by local cadres who had a complex relationship with their community and the socialist state. It triggered a variety of transformations, including a new (relative) prosperity for the rural collectives and a new organisation of production and of rural life more generally.

At last, with the cotton scandal, it became clear that the promise of progress entailed by the Soviet modernisation policy could not be fulfilled for the rural world. Envisaging a 'civilisational advancement' of the countryside, Soviet policy imposed large-scale farms and an agro-industrial bureaucracy on rural people with the idea of rendering rural society tame for the Soviet vision of agricultural modernity. As a result, an inflated service sector created fertile soil for a rural counter-programme against the centre's policies. Hidden behind the new bureaucratic façade, a clientelist mechanism, nominally still inspired by a traditionalistic ethos and supposedly dedicated to the safeguarding of community values, subverted the Soviet modernisation plans from within.

Under Soviet modernisation, rural Uzbek society underwent a substantial mutation. Its leaders and their roles in the communities, the form and substance of family relations, and ways of conceiving and living within community were no longer traditional in the way tsarist-period observers and early Soviet ethnographers had described (chapter 2). Although in the eyes of the centre, rural society and its elites retained many characteristics of traditional society, those elites evolved far away from the traditional social archetypes they invoked in their centre-periphery struggle. Therefore, what consolidated into a deviant system later disclosed by the cotton scandal should not be interpreted as a form of rural resistance, a sort of new *basmachestvo* against Moscow.[155] Although the Central Asian elites mostly

---

[155] The latter was implied by Carlisle (1991: 119) when he wrote that 'turmoil and disarray comparable to the 1984–1989 phase had appeared in the Uzbek SSR during the 1930s', thus associating the political instablity of the 1980s in Soviet Uzbekistan with the Basmachi movement, an uprising by Central Asian Muslim communities against Soviet Russian rule in the 1920s and 1930s.

managed to embody the roles of devout Muslim and loyal communist simultaneously, without contradiction (Khalid 2007: 85), the 'cotton barons' (Rumer 1989) were not against the Soviet order but for it (chapter 3).

Since decollectivisation, the terms of the centre-periphery question have clearly changed (chapter 6). While the cotton economy and its bureaucracy continue to sit at the heart of the power game between centre and periphery, this game now develops in new directions. How the actors, antagonisms, and terms of reference between centre and periphery of the past 'battle for cotton' differ from today's can be illustrated by the crucial conflict that in the past developed around diverging understandings of what was to be a good communist leader. Tellingly, the fundamental conflict today is not over what a good post-communist leader should be. Rather, and more straightforwardly, control over and usufruct in land and assets, and the entrepreneurial liberties granted with them, lie at the centre of the dispute between local elites, rural producers, and the local face of a state that still sometimes leans on the communitarian-kolkhozian ethos. It mostly does so, however, for instrumental reasons, little different from the kolkhoz bosses of the past when they instrumentally resorted to 'tradition' to strengthen their power bases.

Assessing the salience for Central Asia of Gellner's theories about nation-building and Islam, Abashin (2005) concluded that modernisation in Central Asia, contra Gellner (1981, 1983), did not bring long-lasting cultural or social standardisation. Neither agrarian nor industrial, Central Asia's profile, according to Abashin, was that of an intermediate 'agrarian-industrial type', a form of complex society that was remarkably stable and coherent with its underlying economic system. In his opinion, this pre-empted the rise of a genuine modernisation in the name of nationalism and resulted in an unresolved deficit of cultural legitimacy on the part of the leadership, whether the ruling one or a potentially incumbent one:

> In my view, Gellner's prognosis of the necessary 'victory' of industrial over agrarian culture and of nationalism over 'village' Islam is over optimistic. In Central Asia, rather than industrial society what has emerged is a post-colonial 'agrarian-industrial society', which is absent in Gellner's schema. This society, which is based on plantation-kolkhoz (collective farm) cotton production and massive labour migration, is capable of existing for a long time, as it has no possibility of achieving a break-through into the industrial future. ... The leaders of the present Central Asian states are investing enormous energy in order to create and establish feelings of national belonging among the population. However, this attempt arouses strong opposition in their complexly structured societies, giving rise to a justifi-

able suspicion that the nationalism of the elites, like the fundamentalism of the contra-elites, is in fact only a cover for mercenary clan interests (Abashin 2005: 81).

Is the emerging picture another scenario of 'modernisation without development', parallel, for instance, to that of the Mediterranean world (Giordano 1992)? In Mediterranean societies, multiple forms of agrarian protest (Giordano 1992: 83ff.) were a distinctive characteristic accompanying the modernisation path. Along the Soviet modernisation path, violence affected Central Asian rural societies earlier than in the Mediterranean and in the far more virulent form of Stalinist collectivisation, but in doing so it answered much more pervasively the Soviet Union's 'agrarian question' with the 'total social institution' represented by the kolkhoz. Today, after the end of the kolkhoz, the question arises whether Central Asian agrarian-industrial societies will really remain so stable and coherent with their economies as Abashin posited.

Table 4. Relations of power and production in Khorezm before and after decollectivisation.

Variable	Kolkhoz/*Shirkat* System	After Decollectivisation
Management structure	'Black box'	More transparency
Power and organisation of production	Decentralised power; centralised production	Centralised power; decentralised production
Elites and the rule of law	Theft/corruption; illegality	New legal culture; 'legal' theft
Community discourse	Rhetoric of solidarity; 'social cohesion'	Polarisation between winners and losers; 'social desegregation'

Table 4 summarises the most important transformations in relations of power and production in Khorezm before and after the end of the kolkhoz. Post-Soviet agricultural reform measures, instead of focussing on rights and ownership, privileged the restructuring of management and the enhancement of productivity. The collectives of the past were 'black boxes' to the central authorities, in that the centre's capacity to control and steer local processes was limited to the appointment of intermediaries from among the communities' elites, who succeeded in keeping outside control relatively feeble. Decollectivisation is now bringing greater transparency to relations of power and production, which therefore become more visible to the power holders at the centre.

So far, decollectivisation in Uzbekistan has differed from that in most other postsocialist countries. It has not been the structural transformation desired by the international community but rather a new repartition of competences and powers along the lines of an essentially preserved command hierarchy. Responsibilities for production have been moving 'down the ladder' (decentralisation of production) while the central command has been strengthened in its monitoring and supervision of agriculture by the transfer of such powers from the collectives to the district administration (centralisation of power).

In this new context, the law has proved to be an ambiguous instrument of governance that local governments use to ensure their grip on agriculture. Instead of granting justice, the new 'rule of law' unfolds as a distorted instrument of control over agricultural producers. As in the past, communal unity and solidarity are evoked and celebrated, but now, with rural society falling apart into new social stratifications and with a government unwilling or unable to mitigate the rising inequalities, they are perceived differently, and new tensions and divergences within rural communities come to the surface. So far the Uzbek authorities, in Yangibozor and elsewhere, have tried to reduce the propensity of the decollectivising communities to fall apart in new divisions and egoisms, but it remains to be seen how this strategy will fare in the long term, especially now that divisions between fermers and dehqons have become clearer because of farm consolidation. In the meantime, even if capitalist relations have been introduced in Uzbekistan in a filtered way, venality has taken over new spaces, and the market principle seems to have gained territory even in the presence of illiberal conditions.

The tensions accompanying the transition from collective to decollectivised agriculture also have repercussions on the realm of kinship. After the kolkhoz, the Khorezmian rural family seems to be undergoing 'repatriarchalisation', and the social concomitants of Soviet modernisation seem to be giving way dramatically to a return to the individual or 'private' farming (*yakka xo'jalik*) described by Sazonova for the pre-revolutionary period (chapter 2). In Yangizobor the farms of Maxmud, Safarboy, and Tojiboy, described in chapter 5, strikingly resemble, in their structure and internal organisation, those of the Maxsum and Po'stak families that Sazonova described, despite the 70 years of collective socialist agriculture separating them. This process is similar to what Verdery (2003: 348) pointed to regarding decollectivisation in Romania, where the 'retraditionalised peasant' was also emerging as a result of an agrarian reform policy.

I am not arguing that rural society in Khorezm is reverting *en bloc* to pre-Soviet ('traditional') community relations, as if the clock has been turned back. On the contrary, I believe the 'community' celebrated in Soviet kolkhoz rhetoric and promoted by today's authorities with dubious success is falling apart in new divisions and stratifications based on the unequal repartition of risks, opportunities, and liabilities entailed in decollectivisation. Consequently, the concept of re-patriarchalisation should suggest a return neither to past community forms nor to past leadership patterns. Rather, the reverse development can be observed: the role of kinship in securing livelihoods is being strengthened even as the role of kinship in political networking and patronage relations is diminishing because Khorezmian 'big men' must become more individualistic (political) entrepreneurs in order to be successful.[156]

In Khorezm, then, in the transition from collective to decollectivised agriculture, the role of kinship is ambivalent. Kinship relations and networks, often assumed to be the privileged pattern of interaction in postsocialist informal politics, deserve to be approached with greater caution than has so far been the case in the literature, and they demand a broader basis of evidence.[157]

In Khorezm, the competitive 'sponsoring' of farms by emerging, urban-based 'new farmers' (*yangi fermerlar*) and by new-but-old rural notables, the 'large farmers' (*katta fermerlar*), has supplanted the certainty and securities of the patron-client relations of the pre-reform collectives, which were centred on the figure of the *rais*. Since decollectivisation, the larger networks of 'organisational brokerage' (Roniger 1990) that were strong during the growth of the Soviet agricultural bureaucracy have given way to more erratic and fragmented local forms of patronage by individualistic 'political entrepreneurs'. Such entrepreneurs must rely increasingly on their bureaucratic capital in order to stay on top in agriculture, but entrepreneurial patronage relations also emerge as a way of circumventing or even challenging an overwhelming state authority.

In the late 1990s, the 'Uzbek model' of agricultural reform (Ilkhamov 1998) presented a mix of elements of both command and market economies, featuring a number of restrictive characteristics aimed at keeping social and economic processes in rural areas within a track designated by the govern-

---

[156] On the basis of an analysis of the career paths of former shirkat chairmen in two districts of Khorezm, Wegerich (2006: 126) reached a similar conclusion: 'Money as transactional content becomes more important than family and friendship ties'.

[157] Following Leach (1961: 83), one could here say that 'the kinship pattern is not the cause of what happens, it merely provides a framework of ideas in terms of which actual behaviour may be justified. ... it does not finally determine what men do'.

ment. Unlike the more successful Chinese model, in which small producers enjoyed freedoms that boosted productivity within a framework of state predominance[158], in Uzbekistan the mix of reforms and continuance of state regulations did not allow the development of an analogous dynamism in the private farming sector (Pomfret 2000).

Nevertheless, with decollectivisation some invigorating of the agricultural sector, which had been contained by the maintenance of the shirkats, has begun. Although it remains to be seen whether decollectivisation will boost productivity and revitalise the stagnating rural economy, as predicted by the government, its first result has been that a distinct class of agricultural entrepreneurs, with strengthened interests and a sense of ownership, has taken over the burden of state crop production in Uzbekistan. Even if, in the immediate aftermath of decollectivisation, fermers still had no full ownership of land, but merely held leases, they acknowledged that ownership would probably become available in the future. Therefore, despite of the maintenance of constraints and state interference, being a fermer was desirable because of the awareness that only for fermers would there be a future in agriculture.

With decollectivisation, the redefinition of political, economic, and social relations in rural areas of Uzbekistan has begun. The government's agrarian development vision suggested that a growing number of rural people would be pushed out of the agricultural sector, existing farms would grow in size, and the number of farms would diminish, all of which have been largely confirmed by the turn of events of 2008–2009, following from the policy of farm consolidation. With the sizes of farms growing and the number of fermers necessarily decreasing, the generation of fermers who were pushed into agriculture with the dismantling of the shirkats have paid a disproportionately heavy price for the modernisation of agriculture. It was they who carried the burden of enhancing agriculture and exposure to economic risks, even while their prospects for maintaining their leases in the future were uncertain.

Another consideration regarding the transformation of the rural sector is that of the future role of the government. Contrary to fermers' expectations of private ownership of land, loosening the regulatory framework of agriculture or even introducing full ownership of land would represent a major hazard to the government's capacity to contain threatening social developments, such as the emergence of an autonomous, economically

---

[158] In China the preconditions for the success of decollectivisation lay not in the strengthening of private farming and ownership entitlements but in changes in relative prices and procurement policies that turned more favourable for agricultural producers (Anderson 2010: 83; Bramall 1993).

capable, and politically conscious class of rural elites. Despite the government's expressed commitment to greater market accessibility and ownership rights, the future path of agrarian transition remains uncertain to outsiders and highly dependent on the decisions of the national leadership.

By adopting its unprecedented path of agrarian transition, the Uzbek government appears to be seeking a variation of the model of modernising China, in which the gradual strengthening of market reforms follows the goal of economic growth and preservation of an essentially authoritarian state. Wang (1994: 201), writing about the evolution of the Taiwanese political system, stated that 'clientelism and factionalism do not necessarily diminish in importance with economic development', and 'it is possible for a regime to strike a satisfactory equilibrium between clientelism and market economy'. In Uzbekistan the wish to achieve this kind of equilibrium seems to underlie much of the government's gradualist reform pattern. In rural Uzbekistan this kind of balance must revolve around the relationship between state, communities, and producers. Yet on the local level, until the Uzbek modernisation programme delivers on its promise of jobs and wealth, the legitimacy question will remain open and undermine this carefully held balance.

Appendix

Table 5
Cotton harvests on Yangibozor collective farms in the 1960s (in quintals per hectare)

Collective Farm	1960	1961	1962	1963	1964	1965	1966	1967	1968	1969
Bog'olon	22.1	24.4	24.4	25.2	23.7	31.0	32.2	32.1	35.7	27.8
Bo'ston	NE	NE	NE	NE	NE	NE	NE	—	—	23.2
Bo'zqal'a	22.1	20.0	19.7	23.5	25.1	31.0	32.7	33.4	35.9	30.3
Xalqobod	23.4	21.6	20.0	22.6	21.6	28.1	29.5	28.2	33.4	28.7
Jayhun	NE	NE	NE	NE	NE	NE	NE	NE	NE	NE
Hamza	26.1	28.5	27.3	30.1	29.5	32.1	33.9	31.7	36.5	29.1
Xorazm	27.3	25.0	25.9	27.4	23.8	28.3	30.0	24.7	34.2	26.7
Madaniyat	22.6	23.7	22.8	25.0	24.1	31.2	35.0	32.2	36.3	26.4
Sanjar	NE	NE	NE	NE	NE	NE	NE	NE	NE	NE
Shiringo'ng'irot	23.0	21.0	21.0	26.0	23.0	28.0	28.0	29.0	31.0	27.0
O'zbekiston	23.2	25.3	25.2	29.0	26.6	NA	32.4	30.2	35.1	28.4
Average	23.7	23.7	23.3	26.1	24.7	26.2	31.7	30.1	34.7	27.5

Source: Khorezm State Archive, Yangibozor District Branch.
Key: NE, enterprise nonexistent at the time; NA, data unavailable; —, no cotton production.

Table 6
Area planted in cotton on Yangibozor collective farms in the 1960s (in hectares)

Farm	1960	1961	1962	1963	1964	1965	1966	1967	1968	1969
Bog'olon	1,513	1,570	1,570	1,575	1,705	1,678	1,698	1,710	1,750	1,780
Bo'ston	NE	NE	NE	NE	NE	NE	NE	—	—	100
Bo'zqal'a	1,310	1,350	1,340	1,342	1,277	905	905	920	945	975
Xalqobod	1,185	1,211	1,220	1,223	1,306	1,223	1,225	1,185	1,185	1,205
Jayhun	NE	NE	NE	NE	NE	NE	NE	NE	NE	NE
Hamza	680	680	695	692	709	810	810	678	726	756
Xorazm	625	660	675	678	790	771	760	776	790	810
Madaniyat	832	871	870	872	929	810	750	765	795	825
Sanjar	NE	NE	NE	NE	NE	NE	NE	NE	NE	NE
Shiringo'ng'irot	735	759	760	762	830	1,287	1,266	1,280	1,300	1,320
O'zbekiston	1,145	1,140	1,140	1,143	1,151	NA	1,223	1,195	1,210	1,220
Total	8,025	8,141	8,270	8,287	8,697	7,484	8,637	8,509	8,701	8,991

Source: Khorezm State Archive, Yangibozor District Branch.
Key: NE, enterprise nonexistent at the time; NA, data unavailable; —, no cotton production.

Table 7
Cotton harvests on Yangibozor collective farms in the 1970s (in quintals per hectare)

Farm	1970	1971	1972	1973	1974	1975	1976	1977	1978	1979
Bog'olom	40.6	38.7	31.2	38.7	38.6	38.7	31.6	35.7	26.4	37.9
Bo'ston	—	—	—	—	—	—	—	—	—	—
Bo'zqal'a	39.7	39.5	34.8	35.9	40.9	41.9	38.7	40.4	35.8	46.3
Xalqobod	38.7	37.8	30.9	32.7	38.7	37.7	31.5	37.6	28.5	38.5
Jayhun	NE	NE	NE	NE	NE	NE	—	—	—	—
Hamza	38.2	34.1	26.8	34.5	36.7	40.7	34.2	37.8	26.8	40.6
Xorazm	37.6	32.3	29.7	34.2	36.0	37.0	32.0	33.6	29.7	37.2
Madaniyat	39.02	35.3	33.5	34.8	36.1	38.8	36.4	42.0	32.1	39.1
Sanjar	NE	NE	NE	NE	NE	NE	NE	NE	NE	NE
Shirinqo'ng'irot	36.0	33.0	28.0	33.0	35.0	36.0	33.0	34.0	32.0	38.0
O'zbekiston	36.5	33.4	26.9	32.3	38.6	40.7	33.8	39.5	28.2	NA
Average	38.3	35.5	30.2	34.5	37.6	38.9	33.9	37.6	29.9	39.7

Source: Khorezm State Archive, Yangibozor District Branch.
Key: NE, enterprise nonexistent at the time; NA, data unavailable; —, no cotton production.

Table 8
Area planted in cotton on Yangibozor collective farms in the 1970s (in hectares)

Farm	1970	1971	1972	1973	1974	1975	1976	1977	1978	1979
Bog'olon	1,803	1,926	1,820	1,840	1,890	1,800	1,800	1,856	1,845	1,845
Bo'ston	—	—	—	—	—	—	—	—	—	—
Bo'zqal'a	979	1,038	1,000	1,002	1,015	1,021	1,021	1,046	1,046	1,046
Xalqobod	1,208	1,265	1,242	1,225	1,217	1,212	1,210	1,220	1,220	1,220
Jayhun	NE	NE	NE	NE	NE	NE	—	—	—	—
Hamza	787	884	830	840	875	881	881	891	891	891
Xorazm	815	909	860	865	880	886	886	870	840	840
Madaniyat	843	932	870	880	890	896	896	916	916	916
Sanjar	NE	NE	NE	NE	NE	NE	NE	NE	NE	NE
Shirinqo'ng'irot	1,334	1,359	1,340	1,343	1,342	1,351	1,351	1,381	1,370	1,370
O'zbekiston	1,265	1,274	1,200	1,200	1,210	1,266	1,257	1,272	1,272	NA
Total	9,034	9,587	9,162	9,195	9,319	9,304	9,302	9,452	9,400	8,128

Source: Khorezm State Archive, Yangibozor District Branch.
Key: NE, enterprise nonexistent at the time; NA, data unavailable; —, no cotton production.

Table 9
Cotton harvests on Yangibozor collective farms in the 1980s (in quintals per hectare)

Farm	1980	1981	1982	1983	1984	1985	1986	1987	1988	1989
Bog'olon	38.8	NA	NA	NA	NA	NA	NA	NA	NA	NA
Bo'ston	—	—	—	—	—	—	—	—	—	—
Bo'zqal'a	45.7	44.3	45.4	47.6	36.1	37.5	29.6	27.1	32.3	30.1
Xalqobod	38.9	37.9	38.1	40.8	29.0	33.3	22.6	27.6	29.3	45.1
Jayhun	—	—	—	—	—	—	—	—	27.7	27.8
Hamza	40.5	38.2	37.8	41.4	32.4	35.1	27.1	24.5	31.9	29.2
Xorazm	36.7	32.5	35.9	41.5	34.1	40.6	35.1	28.0	30.8	30.0
Madaniyat	42.0	38.7	37.9	41.5	32.5	35.5	30.5	27.9	30.0	30.7
Sanjar	NE	NE	NE	—	—	—	—	—	—	—
Shirinqo'ng'irot	40.0	32.0	36.0	40.0	29.0	33.0	26.0	25.0	32.0	58.0
O'zbekiston	38.4	33.9	39.0	41.1	34.9	34.7	28.0	26.8	31.6	29.3
Average	40.1	36.7	38.6	42.0	32.6	35.7	28.4	26.7	30.7	29.3

Source: Khorezm State Archive, Yangibozor District Branch.
Key: NE, enterprise nonexistent at the time; NA, data unavailable; —, no cotton production.

Table 10
Area planted in cotton on Yangibozor collective farms in the 1980s (in hectares)

Farm	1980	1981	1982	1983	1984	1985	1986	1987	1988	1989
Bog'olon	1,845	NA	NA	NA	NA	NA	NA	NA	NA	NA
Bo'ston	—	—	—	—	—	—	—	—	—	—
Bo'zqal'a	1,046	1,050	1,070	1,097	1,264	1,190	1,262	1,372	1,271	1,270
Xalqobod	1,230	1,230	1,240	1,072	1,427	1,183	1,426	1,509	1,500	1,320
Jayhun	—	—	—	—	—	—	—	—	50	217
Hamza	891	892	897	917	1,010	997	1,083	1,179	1,095	1,099
Xorazm	840	840	835	845	932	910	940	1,068	1,019	1,000
Madaniyat	916	918	938	977	1,132	1,080	1,207	1,312	1,275	1,230
Sanjar	NE	NE	NE	—	—	—	—	—	—	—
Shirinqo'ng'irot	1,370	1,368	1,378	1,410	1,626	1,519	1,550	1,783	1,662	1,560
O'zbekiston	1,272	1,274	1,284	1,316	1,469	1,376	1,436	1,596	1,570	1,500
Total	9,410	7,572	7,642	7,634	8,860	8,255	8,904	9,819	9,442	9,196

Source: Khorezm State Archive, Yangibozor District Branch.
Key: NE, enterprise nonexistent at the time; NA, data unavailable; —, no cotton production.

Bibliography

Abashin, S. 2005. Gellner, the "Saints" and Central Asia: Between Islam and Nationalism. Translated by C. Humphrey. *Inner Asia* 7: 65–86.
——. 2006. The Logic of Islamic Practice: A Religious Conflict in Central Asia. *Central Asian Survey* 25 (3): 267–286.
——. 2007. *Die Sartenproblematik in der Russischen Geschichtsschreibung des 19. und des ersten Viertels des 20. Jahrhunderts*. Berlin: Klaus Schwarz Verlag.
Akiner, S. 1998. Social and Political Reorganization in Central Asia: Transition from Pre-colonial to Post-colonial Society. In T. Atabaki, and J. O'Kane (eds.), *Post-Soviet Central Asia*, pp. 1–34. London: Tauris.
Allina-Pisano, J. 2008. *The Post-Soviet Potemkin Village: Politics and Property Rights in the Black Earth*. New York: Cambridge University Press.
Allworth, E. A. 1990. *The Modern Uzbeks, from the Fourteenth Century to the Present: A Cultural History*. Stanford, California: Hoover Institution Press.
Anderson, D. G., and F. Pine. 1995. Surviving the Transition: Development Concerns in the Post-Socialist World. *Special issue of Cambridge Anthropology* 18 (2).
Anderson, J. 1999. *Kyrgyzstan: Central Asia's Island of Democracy?* Amsterdam: Harwood Academic.
Anderson, P. 2010. Two Revolutions. *New Left Review* 61: 59–96.
Aw-Hassan, A., L. Iniguez, M. Musaeva, M. Suleimenov, R. Khusanov, B. Moldashev, S. Kherremov, A. Ajibekov, and Y. Yakhshilikov. 2004. Economic Transition Impact on Livestock Production in Central Asia: Survey Results. In J. Ryan, P. Vlek, and R. Paroda (eds.), *Agriculture in Central Asia: Research for Development*, pp. 302-350. Bonn and Aleppo: ZEF and ICARDA.
Bailey, F. G. 2001 [1969]. *Stratagems and Spoils: A Social Anthropology of Politics*. Boulder, Colorado: Westview Press.
Baldauf, I. 1995. Identitätsmodelle, Nationenbildung und regionale Kooperation in Mittelasien. In B. Staiger (ed.), *Nationalismus und regionale Kooperation in Asien*, pp. 21–57. Hamburg: Institut für Asienkunde.
Banton, M. (ed.). 1966. *The Social Anthropology of Complex Societies*. London: Tavistock.
Barfield, T. J. 1990. Tribe and State Relations: The Inner Asian Perspective. In P. S. Khoury, and J. Kostiner (eds.), *Tribes and State Formation in the Middle East*, pp. 153–182. Berkeley: University of California Press.

Baskakov, N. A. 1960. *Tyurkskie yazyki*. Moscow: Izdatel'stvo vostochnoy literatury.
Battersby, H. 1969. A Survey of Uzbek Settlement, with Regard to Some Economic and Shelter Changes in the Khwarizm (Khiva) Oasis, North to the Delta of the Amu Darya, Based on Russian Ethnographic Reports. *Tarih Araştırmaları Dergisi* 7: 17–32.
Bayat, A. 1997. *Street Politics: Poor People's Movements in Iran, 1977–1990*. New York: Columbia University Press.
Bellér-Hann, I., and C. Hann. 1999. Peasants and Officials in Southern Xinjiang: Subsistence, Supervision and Subversion. *Zeitschrift für Ethnologie* 124: 1–32.
Benningsen, A., and S. E. Wimbush. 1985. *Muslims of the Soviet Empire: A Guide*. London: C. Hurst.
Bernard, H. R. 1995. *Research Methods in Anthropology: Qualitative and Quantitative Approaches*. Second edition. Walnut Creek, California: Altamira Press.
Bierschenk, T. 1988. Development Projects as Arenas of Negotiation for Strategic Groups: A Case Study from Benin. *Sociologia Ruralis* 38: 147–160.
——, and G. Elwert (eds.). 1993. *Entwicklungshilfe und ihre Folgen: Ergebnisse empirischer Untersuchungen in Afrika*. Frankfurt: Campus Verlag.
——, and J. P. Olivier de Sardan. 1997. ECRIS: Rapid Collective Inquiry for the Identification of Conflicts and Strategic Groups. *Human Organization* 56 (2): 238–244.
Bikzhanova, M. A., K. L. Zadykhina, and O. A. Sukhareva. 1974. Social and Family Life of the Uzbeks. In E. Dunn, and S. P. Dunn (eds.), *Introduction to Soviet Ethnography*, pp. 239–271. Berkeley, California: Highgate Road Social Science Research Station.
Blum, A., and M. Mespoulet. 2003. *L'Anarchie Bureaucratique: Statistique et Pouvoir sous Staline*. Paris: La Découverte.
Boissevain, J. 1974. *Friend of Friends: Networks, Manipulators and Coalitions*. Oxford: Basil Blackwell.
Bourdieu, P. 1985. The Forms of Capital. In J. G. Richardson (ed.), *The Handbook of Theory and Research for the Sociology of Education*, pp. 241–258. New York: Greenwood.
Bramall, C. 1993. The Role of Decollectivisation in China's Agricultural Miracle. *Journal of Peasant Studies* 20 (2): 271–295.
Brandtstätter, S. 2003. The Moral Economy of Kinship and Poverty in Southern China. In C. Hann and the Property Relations Group (eds.),

The Postsocialist Agrarian Question: Property Relations and the Rural Condition, pp. 419–440. Münster: LIT Verlag.

Bregel, Y. 1986. Qosh-begi. In C. E. Bosworth, E. van Donzel, B. Lewis, and C. Pellat (eds.), *The Encyclopaedia of Islam*, new edition, vol. 5, pp. 273–274. Leiden: E. J. Brill.

——. 2004. Atalïq. In P. J. Bearman, T. Bianquis, C. E. Bosworth, E. van Donzel, and W. P. Heinrichs (eds.), *The Encyclopaedia of Islam*, new edition, supplement, vol. 12, pp. 96–98. Leiden: Koninklijke Brill.

Brüder Grimm. 1812. Die Goldene Gans. In *Kinder- und Hausmärchen*, vol. 1, pp. 303–308. Berlin: Realschulbuchhandlung.

Burawoy, M., and K. Verdery (eds.). 1999. *Uncertain Transition: Ethnographies of Change in the Postsocialist World*. Lanham, Maryland: Rowman and Littlefield.

Buttino, M. 2004. Dopo la fine del regime sovietico: Il caso uzbeco. In P. Viola, and A. Blando (eds.), *Quando crollano i regimi*, pp. 180–200. Palermo: Palumbo.

Carlisle, D. S. 1991. Power and Politics in Soviet Uzbekistan: From Stalin to Gorbachev. In W. Fierman (ed.), *Central Asia: The Failed Transformation*, pp. 93–129. Boulder, Colorado: Westview Press.

Cartwright, A. L. 2001. *The Return of the Peasant: Land Reform in Post-Communist Romania*. Aldershot, UK: Ashgate/Dartmouth.

Cellarius, B. A. 2003. Property Restitution and Natural Resource Use in the Rhodope Mountains, Bulgaria. In C. Hann and the Property Relations Group (eds.), *The Postsocialist Agrarian Question: Property Relations and the Rural Condition*, pp. 189–218. Münster: LIT Verlag.

Christophe, B. 2005. *Metamorphosen des Leviathan in einer postsozialistischen Gesellschaft: Georgiens Provinz zwischen Fassaden der Anarchie und regulativer Allmacht*. Bielefeld: transcript Verlag.

Chylinski, E. 1991. Ritualism of Family Life in Soviet Central Asia: The Sunnat (Circumcision). In S. Akiner (ed.), *Cultural Change and Continuity in Central Asia*, pp. 160–170. London and New York: Kegan Paul International and Central Asia Research Forum.

Collins, K. 2004. The Logic of Clan Politics: Evidence from the Central Asian Trajectories. *World Politics* 56: 224–261.

Conquest, R. 1986. *The Harvest of Sorrow: Soviet Collectivization and the Terror Famine*. Oxford: Oxford University Press.

——. 1991. *Stalin: Breaker of Nations*. New York: Penguin Books.

Creed, G. 1999. Deconstructing Socialism in Bulgaria. In M. Burawoy, and K. Verdery (eds.), *Uncertain Transition: Ethnographies of Change in the Postsocialist World*, pp. 223–243. Lanham, Maryland: Rowman and Littlefield.

Critchlow, J. 1991. Prelude to 'Independence': How the Uzbek Party Apparatus Broke Moscow's Grip on Elite Recruiting. In W. Fierman (ed.), *Central Asia: The Failed Transformation*, pp. 131–156. Boulder, Colorado: Westview Press.

Defrade, D. 2000. *La decollectivisation des campagnes ouzbèques: Vers une transformation des modes d'exploitation?* DEA thesis, EHESS, Paris.

Djalalov, S. 2007. Indirect Taxation of the Uzbek Cotton Sector: Estimation and Policy Consequences. In D. Kandiyoti (ed.), *The Cotton Sector in Central Asia: Economic Policy and Development Challenges*, pp. 90–101. London: School of Oriental and African Studies, University of London.

Djanibekov, N. 2008a. *A Micro-economic Analysis of Farm Restructuring in the Khorezm Region*. PhD dissertation, Zentrum für Entwicklungsforschung, University of Bonn.

———. 2008b. A Model-based Analysis of Land and Water Use Reforms in Khorezm: Effects on Different Types of Agricultural Producers. In P. Wehrheim with A. Schoeller-Schletter, and C. Martius (eds.), *Continuity and Change: Land and Water Use Reforms in Rural Uzbekistan*, pp. 43–62. Halle: Leibniz-Institut für Agrarentwicklung in Mittel- und Osteuropa (IAMO).

Dragadze, T. 1984. Introduction. In T. Dragadze (ed.), *Family and Kinship in the Soviet Union*, pp. 1–9. London: Routledge and Kegan Paul.

Dumont, R. 1964. *Sovkhoz, Kolkhoz, ou le problématique communisme*. Paris: Seuil.

Eckert, J. 1996. *Das unabhängige Usbekistan: Auf dem Weg von Marx zu Timur. Politische Strategien der Konfliktregelung in einem Vielvölkerstaat*. Münster: LIT Verlag.

——— (ed.). 2004. *Anthropologie der Konflikte: Georg Elwerts konflikttheoretische Thesen in der Diskussion*. Bielefeld: transcript Verlag.

———, and G. Elwert. 2000. *Land Tenure in Uzbekistan*. Sozialanthropologische Arbeitspapiere 86. Berlin: Das Arabische Buch Verlag.

Eickelman, D. F. 2002. *The Middle East and Central Asia: An Anthropological Approach*. Fourth edition. Upper Saddle River, New Jersey: Prentice Hall.

Eisener, R. 1994. Some Problems of Research concerning the National Delimitation of Soviet Central Asia in 1924. In B. G. Fragner, and B. Hoffmann (eds.), *Bamberger Mittelasienstudien*, pp. 109–116. Berlin: Klaus Schwarz Verlag.

Eisenstadt, S. N., and L. Roniger. 1980. Patron-Client Relations as a Model of Structuring Social Exchange. *Comparative Studies in Society and History* 22 (1): 42–77.
——. 1984. *Patrons, Clients and Friends*. Cambridge: Cambridge University Press.
Elwert, G. 1983. *Bauern und Staat in Westafrika: Die Verflechtung sozioökonomischer Sektoren am Beispiel Bénin*. Frankfurt a. M.: Campus Verlag.
——. 1989. Nationalismus und Ethnizität: Über die Bildung von Wir-Gruppen. *Kölner Zeitschrift für Soziologie und Sozialpsychologie* 3: 440–464.
——. 1995. Boundaries, Cohesion and Switching: On We-Groups in Ethnic, National and Religious Form. In B. Brumen, and Z. Šmitek (eds.), *Bulletin of the Slovene Ethnological Society* (*Mediterranean Ethnological Summer School*) 24: 105–121.
——. 2001. Conflict: Anthropological Aspects. In N. Smelser, and P. Baltes (eds.), *International Encyclopedia of the Social and Behavioral Sciences*, pp. 2542–2547. Amsterdam: Elsevier Science.
——. 2003. *Feldforschung: Orientierungswissen und kreuzperspektivische Analyse*. Sozialanthropologische Arbeitspapiere 96. Berlin: Das Arabische Buch Verlag.
Evers, H.-D., and T. Schiel. 1988. *Strategische Gruppen: Vergleichende Studien zu Staat, Bürokratie und Klassenbildung in der Dritten Welt*. Berlin: Reimer Verlag.
Fathi, H. 2004. Islamisme et pauvreté dans le monde rural de l'Asie centrale post-soviétique: Vers un espace de solidarité islamique? *UNRISD Civil Society and Social Movements Paper* 14. Geneva: UNRISD.
Fedtke, G. 2007. How Bukharans Turned into Uzbeks and Tajiks: Soviet Nationalities Policy in the Light of a Personal Rivalry. In P. Sartori, and T. Trevisani (eds.), *Patterns of Transformation in and around Uzbekistan*, pp. 19–50. Reggio Emilia: Diabasis.
Fierman, W. (ed.). 1991. The Soviet 'Transformation' of Central Asia. In W. Fierman (ed.), *Central Asia: The Failed Transformation*, pp. 11–35. Boulder, Colorado: Westview Press.
——. 1997. Political Development in Uzbekistan: Democratization? In K. Dawisha, and B. Parrot (eds.), *Conflict, Cleavage and Change in Central Asia and the Caucasus*, pp. 360–408. Cambridge: Cambridge University Press.
Finke, P. 2002. Retraditionalisierung und gesellschaftliche Transformation. In A. Strasser (ed.), *Zentralasien und Islam*, pp. 137–149. Hamburg: Deutsches Orient-Institut.

———. 2006. *Variations on Uzbek Identity: Concepts, Constraints and Local Configurations*. Habilitation thesis, University of Leipzig and Max Planck Institute for Social Anthropology, Halle.
Fitzpatrick, S. 1994. *Stalin's Peasants: Resistance and Survival in the Russian Village after Collectivization*. Oxford: Oxford University Press.
Foster, G. 1973. *Traditional Societies and Technological Change*. New York: Harper and Row.
Fumagalli, M. 2007. Informal (Ethno-)Politics and Local Authority Figures in Osh, Kyrgyzstan. *Ethnopolitics* 6 (2): 211–233.
Gambold Miller, L., and P. Heady. 2003. Cooperation, Power, and Community: Economy and Ideology in the Russian Countryside. In C. Hann and the Property Relations Group (eds.), *The Postsocialist Agrarian Question: Property Relations and the Rural Condition*, pp. 257–292. Münster: LIT Verlag.
Gdlyan, T., and N. Ivanov. 1994. *Kremlevskoe delo*. Rostov na Donu: AO Kniga.
Geertz, C. 1963. *Agricultural Involution*. Berkeley: University of California Press.
Geiss, P. G. 2003. *Pre-Tsarist and Tsarist Central Asia: Communal Commitment and Political Order in Change*. London: RoutledgeCurzon.
Gellner, E. 1981. *Muslim Society*. Cambridge: Cambridge University Press.
———. 1983. *Nations and Nationalism*. Oxford: Blackwell.
Germanov, V. A. 2002. P. Tolstov: Maître, docteur, commandeur, ou l'histoire à travers l'archéologie et l'ethnographie. *Cahiers d'Asie Centrale* 10: 193–215.
Giordano, C. 1992. *Die Betrogenen der Geschichte: Überlagerungsmentalität und Überlagerungsrationalität in mediterranen Gesellschaften*. Frankfurt a. M.: Campus Verlag.
———, and D. Kostova. 2002. The Social Production of Mistrust. In C. Hann (ed.), *Postsocialism: Ideals, Ideologies and Practices in Eurasia*, pp. 74–94. London: Routledge.
Gleason, G. 1983. The Pakhta Programme: The Politics of Sowing Cotton in Uzbekistan. *Central Asian Survey* 2 (2): 109–120.
———. 1986. Sharaf Rashidov and the Dilemmas of National Leadership. *Central Asian Survey* 5 (3–4): 133–160.
———. 1991. Fealty and Loyalty: Informal Authority Structures in Soviet Asia. *Soviet Studies* 43 (4): 613–628.
———. 2003. *Markets and Politics in Central Asia*. London: Routledge.
Gluckman, M. 1959. Political Institutions. In E. E. Evans-Pritchard (ed.), *The Institutions of Primitive Society: A Series of Broadcast Talks*, pp. 66-80. Oxford: Basil Blackwell.

―――. 1967. Introduction. In A. L. Epstein (ed.), *The Craft of Social Anthropology*, pp. xi–xx. London: Tavistock.

Godelier, M., and M. Strathern (eds.). 1991. *Big Men and Great Men: Personifications of Power in Melanesia*. Cambridge: Cambridge University Press.

Griffin, K., A. R. Khan, and A. Ickowitz. 2002. Poverty and the Distribution of Land. In U. K. Ramachandran, and M. Swaminathan (eds.), *Agrarian Studies: Essays on Agrarian Relations in Less-Developed Countries*, pp. 3–53. New Delhi: Tulika Books.

Guadagni, M., M. Raiser, A. Crole-Rees, and D. Khidirov. 2005. *Cotton Taxation in Uzbekistan: Opportunities for Reform*. ECSSD Working Paper 41. Washington, DC: World Bank.

Guliyamov, Ya. G. 1959. *Xorazmning Sug'orish Tarixi: Qadimgi zamonlardan xozirgacha*. Tashkent: SSR Fanlar Akademiyasi Nashriyoti.

Hann, C. 1993. Introduction: Social Anthropology and Socialism. In C. Hann (ed.), *Socialism: Ideals, Ideologies and Local Practice*, pp. 1–26. London: Routledge.

――― (ed.). 1998. *Property Relations: Renewing the Anthropological Tradition*. Cambridge: Cambridge University Press.

――― (ed.). 2002. *Postsocialism: Ideals, Ideologies and Practices in Eurasia*. London: Routledge.

―――. 2003. Introduction: Decollectivisation and the Moral Economy. In C. Hann and the Property Relations Group (eds.), *The Postsocialist Agrarian Question: Property Relations and the Rural Condition*, pp. 1–46. Münster: LIT Verlag.

―――. 2004. Landwirtschaftsgenossenschaften, Langfristrechte und Legitimation: Eine Fallstudie aus Ungarn. In J. Eckert (ed.), *Anthropologie der Konflikte: Georg Elwerts konflikttheoretische Thesen in der Diskussion*, pp. 217–230. Bielefeld: transcript Verlag.

―――. 2005. Postsocialist Societies. In J. G. Carrier (ed.), *Handbook of Economic Anthropology*, pp. 547–557. Cheltenham: Edward Elgar.

――― and the Property Relations Group (eds.). 2003. *The Postsocialist Agrarian Question: Property Relations and the Rural Condition*. Münster: LIT Verlag.

Harriss, J. 2004. Studying 'Rural Development' Politically. In A. K. Giri (ed.), *Creative Social Research: Rethinking Theories and Methods*, pp. 146–165. New Delhi: Vistaar.

Hilgers, I. 2009. *Why Do Uzbeks Have to Be Muslims? Exploring Religiosity in the Ferghana Valley*. Berlin: LIT Verlag.

Hirschman, A. 1970. *Exit, Voice and Loyalty: Responses to Decline in Firms, Organizations and States.* Cambridge, Massachusetts: Harvard University Press.

Hivon, M. 1995. Local Resistance to Privatization in Rural Russia. In D. Anderson, and F. Pine (eds.), *Surviving the Transition: Development Concerns in the Post-Socialist World.* Special issue of *Cambridge Anthropology* 18 (2): 13–22.

Holt Ruffin, M., and D. C. Waugh (eds.). 1999. *Civil Society in Central Asia.* Seattle: University of Washington Press.

Humphrey, C. 1983. *Karl Marx Collective: Economy, Society and Religion in a Siberian Collective Farm.* Cambridge: Cambridge University Press.

———. 1998. *Marx Went Away but Karl Stayed Behind.* Ann Arbor: University of Michigan Press.

———. 2002. *The Unmaking of Soviet Life: Everyday Economics after Socialism.* Ithaca, New York: Cornell University Press.

Ilkhamov, A. 1998. Shirkats, Dekhqon Farmers and Others: Farm Restructuring in Uzbekistan. *Central Asian Survey* 17 (4): 539–560.

———. 2000. Divided Economy: Kolkhozes vs. Peasant Subsistence Farms in Uzbekistan. *Central Asia Monitor* 4: 5–14.

———. 2001. Impoverishment of the Masses in the Transition Period: Signs of an Emerging "New Poor" Identity in Uzbekistan. *Central Asian Survey* 20 (1): 33–54.

———. 2004. The Limits of Centralization: Regional Challenges in Uzbekistan. In P. Jones Luong (ed.), *The Transformation of Central Asia: States and Societies from Soviet Rule to Independence*, pp. 159–181. Ithaca, New York: Cornell University Press.

———. 2007. National Ideologies and Historical Mythology Construction in Post-Soviet Central Asia. In P. Sartori, and T. Trevisani (eds.), *Patterns of Transformation in and around Uzbekistan*, pp. 91–120. Reggio Emilia: Diabasis.

International Crisis Group. 2004. The Failure of Reform in Uzbekistan: Ways Forward for the International Community. *Asia Report* 76.

———. 2005. Uzbekistan: The Andijon Uprising. *Asia Briefing* 38.

International Monetary Fund. 2000. *Republic of Uzbekistan: Recent Economic Developments.* Washington, DC: International Monetary Fund.

Jones Luong, P. 2002. *Institutional Change and Political Continuity in Post-Soviet Central Asia: Power, Perceptions, and Pacts.* Cambridge: Cambridge University Press.

———. 2004a. Introduction: Politics in the Periphery. Competing Views of Central Asian States and Societies. In P. Jones Luong (ed.), *The Transformation of Central Asia: States and Societies from Soviet Rule to Independence*, pp. 1–26. Ithaca, New York: Cornell University Press.

———. 2004b. Conclusion: Central Asia's Contribution to Theories of the State. In P. Jones Luong (ed.), *The Transformation of Central Asia: States and Societies from Soviet Rule to Independence*, pp. 271–282. Ithaca, New York: Cornell University Press.

Jumaniyozov, M. 2008. *Esimda qolgan onlar*. Tashkent: Musiqa.

Kaxramanov, T. 2000. *Vozvrashchenie iz 'Ada'*. Tashkent: Yozuvchi.

Kamp, M. 2004. Between Women and the State: Mahalla Committee and Social Welfare in Uzbekistan. In P. Jones Luong (ed.), *The Transformation of Central Asia: States and Societies from Soviet Rule to Independence*, pp. 29–58. Ithaca, New York: Cornell University Press.

———. 2006. *The New Woman in Uzbekistan: Islam, Modernity, and Unveiling under Communism*. Seattle: University of Washington Press.

Kandiyoti, D. 1998. Rural Livelihoods and Social Networks in Uzbekistan: Perspectives from Andijan. In D. Kandyoti, and R. Mandel (eds.), *Market Reforms, Social Dislocations and Survival in Post-Soviet Central Asia*. Special Issue of *Central Asian Survey* 17 (4): 561–578.

———. 1999. Poverty in Transition: An Ethnographic Critique of Household Surveys in Post-Soviet Central Asia. *Development and Change* 30: 499–524.

———. 2000. Modernisation without the Market? The Case of the 'Soviet East'. In A. Arce, and N. Long (eds.), *Anthropology, Development and Modernities: Exploring Discourses, Counter-tendencies and Violence*, pp. 52–63. London: Routledge.

———. 2002. How Far Do Analyses of Postsocialism Travel? The Case of Central Asia. In C. Hann (ed.), *Postsocialism: Ideals, Ideologies and Practices in Eurasia*, pp. 238–257. London: Routledge.

———. 2003a. The Cry for Land: Agrarian Reform, Gender, and Land Rights in Uzbekistan. *Journal of Agrarian Change* 3 (1–2): 225–256.

———. 2003b. Pathways of Farm Restructuring in Uzbekistan: Pressures and Outcomes. In M. Spoor (ed.), *Transition, Institutions, and the Rural Sector*, pp. 143–162. Lanham, Maryland: Lexington Books.

———. 2004. Post-Soviet Institutional Design, NGOs and Rural Livelihoods in Uzbekistan. *UNRISD Civil Society and Social Movements Programme Paper* 11. Geneva: UNRISD.

———. 2007. Introduction. In D. Kandiyoti (ed.), *The Cotton Sector in Central Asia: Economic Policy and Development Challenges*, pp. 1–11. London: School of Oriental and African Studies, University of London.

———. 2009. *Invisible to the World? The Dynamics of Forced Child Labour in the Cotton Sector in Uzbekistan*. London: School of Oriental and African Studies, University of London.

———, and R. Mandel (eds.). 1998. Market Reforms, Social Dislocations and Survival in Post-Soviet Central Asia. Special issue of *Central Asian Survey* 17 (4).

Kaneff, D., and L. Yalçın-Heckmann. 2003. Retreat to the Cooperative or to the Household? Agricultural Privatisation in Ukraine and Azerbaijan. In C. Hann and the Property Relations Group (eds.), *The Postsocialist Agrarian Question: Property Relations and the Rural Condition*, pp. 219–256. Münster: LIT Verlag.

Kautsky, K. 1899. *Die Agrarfrage*. Stuttgart: Dietz.

Kehl-Bodrogi, K. 2006. Who Owns the Shrine? Competing Meanings and Authorities at a Pilgrimage Site in Khorezm. *Central Asian Survey* 25 (3): 235–250.

———. 2008. *'Religion Is Not So Strong Here': Muslim Religious Life in Khorezm after Socialism*. Berlin: LIT Verlag.

Keyder, C., and A. Kudat. 2000. Social Dimensions of Agrarian Transformation. In A. Kudat, S. Peabody, and C. Keyder (eds.), *Social Assessment and Agricultural Reform in Central Asia and Turkey*, pp.1–40. World Bank Technical Paper 461. Washington, DC: World Bank.

Khalid, A. 2007. *Islam after Communism: Religion and Politics in Central Asia*. Berkeley: University of California Press.

Khan, A. R., and D. Ghai. 1979. *Collective Agriculture and Rural Development in Soviet Central Asia*. London: Macmillan.

Kienzler, K. 2010. *Improving Nitrogen Use Efficiency and Crop Quality in the Khorezm Region, Uzbekistan*. ZEF Ecology and Development Series no. 72. Bonn: University of Bonn.

Kim, A. 2007. Abandoned by the State: Cotton Production in South Kyrgyzstan. In D. Kandiyoti (ed.), *The Cotton Sector in Central Asia: Economic Policy and Development Challenges*, pp. 119–125. London: School of Oriental and African Studies, University of London.

Kislyakov, N. A. 1954. *Kul'tura i byt tadzhikskogo kolkhoznogo krest'ianstva: Po materialam kolkhoza im. G. M. Malenkova Leninabadskogo raiona Leninabadskoi oblasti Tadzhikskoi SSR*. Trudy Instituta etnografii im. N. N. Miklukho-Maklaia, new series, vol. 24. Moscow: Izdatel'stvo Akademii nauk SSSR.

Kitching, G. 1998. The Revenge of the Peasant? The Collapse of Large-Scale Russian Agriculture and the Role of the Peasant 'Private Plot' in that Collapse, 1991–1997. *Journal of Peasant Studies* 26 (1): 43–81.

Konrád, G., and I. Szelényi (eds.). 1979. *The Intellectuals on the Road to Class Power*. New York: Harcourt Brace Jovanovich.

Kornai, J. 1992. *The Socialist System: The Political Economy of Communism*. Princeton, New Jersey: Princeton University Press.

Koroteyeva, V., and E. Makarova. 1998a. Money and Social Connections in the Soviet and Post-Soviet Uzbek City. In D. Kandyoti, and R. Mandel (eds.), *Market Reforms, Social Dislocations and Survival in Post-Soviet Central Asia*. Special issue of *Central Asian Survey* 17 (4): 579–596.

———. 1998b. The Assertion of Uzbek National Identity: Nativization or State-Building Process? In T. Atabaki, and J. O'Kane (eds.), *Post-Soviet Central Asia*, pp. 137–143. London: Tauris.

Krader, L. 1963. *Peoples of Central Asia*. Bloomington: Indiana University.

Kressel, G. M. 2004. On 'Studying Up': Thoughts on Method. In A. K. Giri (ed.), *Creative Social Research: Rethinking Theories and Methods*, pp. 220–230. New Delhi: Vistaar.

Kudat, A. 2000. Highlights of the Case Studies. In A. Kudat, S. Peabody, and C. Keyder (eds.), *Social Assessment and Agricultural Reform in Central Asia and Turkey*, pp. 88–117. World Bank Technical Paper 461. Washington DC: World Bank.

———, A. Zholdasov, A. Ilkhamov, and J. Bernstein. 1997. Responding to Needs in Uzbekistan's Aral Sea Region. In M. Cernea, and A. Kudat (eds.), *Social Assessment for Better Development: Case Studies in Russia and Central Asia*. Environmentally Sustainable Development Studies and Monograph Series 16. Washington DC: World Bank.

Lampland, M. 2002. The Advantages of Being Collectivized: Cooperative Farm Managers in the Postsocialist Economy. In C. Hann (ed.), *Postsocialism: Ideals, Ideologies and Practices in Eurasia*, pp. 31–56. London: Routledge.

Leach, E. 1961. *Pul Elya: A Village in Ceylon*. Cambridge: Cambridge University Press.

Lerman, Z. 1998. Land Reform in Uzbekistan. In S. Wegren (ed.), *Land Reform in the Former Soviet Union and in Eastern Europe*, pp. 136–161. London: Routledge.

———, C. Csaki, and G. Feder (eds.). 2004. *Agriculture in Transition: Land Policies and Evolving Farm Structures in Post-Soviet Countries*. Lanham, Maryland: Lexington Books.

Leutloff-Grandits, C. 2006. *Claiming Ownership in Postwar Croatia: The Dynamics of Property Relations and Ethnic Conflict in the Knin Region*. Berlin: LIT Verlag.

Levin, T. 1996. *The Hundred Thousand Fools of God: Musical Travel in Central Asia (and Queens, New York)*. Bloomington: Indiana University Press.

Louw, M. 2007. *Everyday Islam in Post-Soviet Central Asia*. London: Routledge.

Lubin, N., and B. Rubin. 1999. *Calming the Ferghana Valley: Development and Dialogue in the Heart of Central Asia*. Report for the Center for Preventive Action. New York: Century Foundation Press.

Mandel, R. 2002. Seeding Civil Society. In C. Hann (ed.), *Postsocialism: Ideals, Ideologies and Practices in Eurasia*, pp. 279–296. London: Routledge.

Manz, B. F. 1994. Historical Background. In B. F. Manz (ed.), *Central Asia in Historical Perspective*, pp. 4–23. Boulder, Colorado: Westview Press.

Massicard, E., and T. Trevisani. 2003. The Uzbek Mahalla: Between State and Society. In T. Everett-Heath (ed.), *Central Asia: Aspects of Transition*, pp. 205–218. London: RoutledgeCurzon.

Matley, I. M. 1967. Agricultural Development. In E. Allworth (ed.), *Central Asia: A Century of Russian Rule*, pp. 266–308. New York: Columbia University Press.

Matniyozov, M. 1997. *Xorazm Tarihi*, vol. 2. Urganch: Izdaltel'stvo 'Xorazm'.

McChesney, R. D. 1996. *Central Asia: Foundations of Change*. Princeton, New Jersey: Darwin Press.

McMann, K. M. 2004. The Civic Realm in Kyrgyzstan: Soviet Economic Legacies and Activists' Expectations. In P. Jones Luong (ed.), *The Transformation of Central Asia: States and Societies from Soviet Rule to Independence*, pp. 213–245. Ithaca, New York: Cornell University Press.

Megoran, N. 2008. Framing Andijon, Narrating the Nation: Islam Karimov's Account of the Events of 13 May 2005. *Central Asian Survey* 27 (1): 15–31.

Micklin, P. 2000. *Managing Water in Central Asia*. London: Royal Institute of International Affairs.

Migdal, J. S. 1988. *Strong Societies and Weak States: State-Society Relations and State Capabilities in the Third World*. Princeton, New Jersey: Princeton University Press.

Müller, M. 2006. *A General Equilibrium Approach to Modeling Water and Land Use Reforms in Uzbekistan*. PhD dissertation, Zentrum für Entwicklungsforschung, University of Bonn.

Nacou, D. 1958. *Du kolkhoze au sovkhoze*. Paris: Les Editions de Minuit.

Nazpary, J. 2002. *Post-Soviet Chaos: Violence and Dispossession in Kazakhstan*. London: Pluto Press.

Nikulin, A. 2003. Kuban *Kolkhoz* between a Holding and a Hacienda: Contradiction of Post-Soviet Rural Development. *Focaal—European Journal of Anthropology* 41: 137–152.

Nolan, R. 2002. *Development Anthropology: Encounters in the Real World*. Boulder, Colorado: Westview Press.

Olivier de Sardan, J. P. 1995. *Anthropologie et développement: Essai en socio-anthropologie du changement social*. Marseille: Editions Apad et Karthala.

O'tamov, J., and R. Qurbonov. 2001. *Fermer xo'jaliklarini tashkil qilishda Xorazm tajribasi*. Tashkent: 'Hayot-GVG'.

O'zbek Sovyet Entsiklopediyasi. 1979. Vol. 12, Xorazm Oblasti, pp. 365–369. Tashkent: O'zSSR Fanlar Akademiyasi.

O'zbek Tilining Izoxli Lug'ati. 1981. Otalik. Vol. 1, p. 548. Moscow: Izdatel'stvo "Russkiy Yazyk".

Pardo, I. 1996. *Managing Existence in Naples: Morality, Action, and Structure*. Cambridge: Cambridge University Press.

Patnaik, A. 1996. *Central Asia: Between Modernity and Tradition*. Delhi: Konark Publishers.

Paul, J. 2010. Recent Monographs on the Social History of Central Asia. *Central Asian Survey* 29 (1): 119–130.

Petric, B. M. 2002. *Pouvoir, don et réseaux en Ouzbékistan post-soviétique*. Paris: Presses Universitaires de France.

Pine, F. 2002. Retreat to the Household? Gendered Domains in Postsocialist Poland. In C. Hann (ed.), *Postsocialism: Ideals, Ideologies and Practices in Eurasia*, pp. 95–113. London: Routledge.

Polanyi, K. 1957. The Economy as Instituted Process. In K. Polanyi, C. Arensberg, and H. Pearson (eds.), *Trade and Markets in the Early Empires*, pp. 243–270. London: Collier Macmillan.

Pomfret, R. 2000. Agrarian Reform in Uzbekistan: Why Has the Chinese Model Failed to Deliver? *Economic Development and Cultural Change* 48 (2): 269–284.

Putnam, R. 1993. *Making Democracy Work: Civic Traditions in Modern Italy*. Princeton, New Jersey: Princeton University Press.

Radniz, S. 2005. Networks, Localism and Mobilization in Aksy, Kyrgyzstan. *Central Asian Survey* 24 (4): 405–424.

Rasanayagam J. 2002a. Spheres of Communal Participation: Placing the State within Local Modes of Interaction in Rural Uzbekistan. *Central Asian Survey* 21 (1): 55–70.

———. 2002b. *The Moral Construction of the State in Uzbekistan: Its Construction within Concepts of Community and Interaction at the Local Level*. PhD dissertation, Cambridge University.

———. 2003. Market, State and Community in Uzbekistan: Reworking the Concept of the Informal Economy. *Working Paper* 59. Halle: Max Planck Institute for Social Anthropology.

———. 2010. *Islam in Post-Soviet Uzbekistan: The Morality of Experience*. Cambridge University Press.

Rasuly-Paleczek, G. 2006. Comparative Perspectives on Central Asia and the Middle East in Social Anthropology and the Social Sciences (Part 2). *Central Eurasian Studies Review* 5 (2): 7–13.

Roniger, L. 1990. *Hierarchy and Trust in Modern Mexico and Brazil*. New York: Praeger.

Rottenburg, R. 2002. *Weit hergeholte Fakten: Eine Parabel der Entwicklungshilfe*. Stuttgart: Lucius and Lucius.

Roy, O. 1997a. *La nouvelle Asie Centrale ou la fabrication des nations*. Paris: Seuil.

———. 1997b. Groupes de solidarité en Asie centrale et en Afghanistan. *Les Annales de l'Autre Islam* 4: 199-215. Paris: INALCO-ERSIM.

———. 1999. Kolkhoz and Civil Society in the Independent States of Central Asia. In M. Holt Ruffin, and D. C. Waugh (eds.), *Civil Society in Central Asia*, pp. 109–121. Seattle: University of Washington Press.

Rumer, B. 1989. *Soviet Central Asia: 'A Tragic Experiment'*. Boston: Unwin Hyman.

———. 1991. Central Asia's Cotton Economy and Its Costs. In W. Fierman (ed.), *Central Asia: The Failed Transformation*, pp. 62–89. Boulder, Colorado: Westview Press.

Rywkin, M. 1985. Power and Ethnicity: Regional and District Party Staffing in Uzbekistan (1983/84). *Central Asian Survey* 4 (1): 3–40.

Sabol, S. 1995. The Creation of Soviet Central Asia: The 1924 National Delimitation. *Central Asian Survey* 14 (2): 225–241.

Sahlins, M. 1963. Poor Man, Rich Man, Big Man, Chief: Political Types in Melanesia and Polynesia. *Comparative Studies in Society and History* 5: 285–303.

Saktanber, A., and A. Özataş-Baykal. 2000. Homeland within Homeland: Women and the Formation of Uzbek National Identity. In F. Acar, and A. Güneş-Ayata (eds.), *Gender and Identity Construction:*

Women of Central Asia, Caucasus and Turkey, pp. 229–248. Leiden: Brill.

Sartori, P. 2010. Introduction: Dealing with States of Property in Modern and Colonial Central Asia. *Central Asian Survey* 29 (1): 1–8.

Sazonova, M. V. 1952. K etnografii uzbekov iuzhnogo Khorezma. In S. P. Tolstov, and T. A. Zhdanko (eds.), *Trudy Khorezmskoy arkheologo-etnograficheskoi ekspeditsii*, vol. 1, pp. 247–318. Moscow: Izdatel'stvo Akademii nauk SSSR.

Schatz, E. 2004. *Modern Clan Politics: The Power of 'Blood' in Kazakhstan and Beyond*. Seattle: University of Washington Press.

Schoeberlein-Engel, J. S. 1994. *Identity in Central Asia: Construction and Contention in the Conceptions of 'Özbek', 'Tâjik', 'Muslim', 'Samarqandi' and Other Groups*. PhD dissertation, Harvard University.

Schoeller-Schletter, A. 2008. Organizing Agricultural Production: Law and Legal Forms in Transition. In P. Wehrheim, A. Schoeller-Schletter, and C. Martius (eds.), *Continuity and Change: Land and Water Use Reforms in Rural Uzbekistan*, pp. 17–40. Halle: Leibniz-Institut für Agrarentwicklung in Mittel- und Osteuropa (IAMO).

Shanin, T. 1989. Soviet Agriculture and Perestroika: Four Models. The Most Urgent Task and the Farthest Shore. *Sociologia Ruralis* 29 (1): 7–22.

Scott, J. 1985. *Weapons of the Weak: Everyday Forms of Peasant Resistance*. New Haven, Connecticut: Yale University Press.

———. 1998. *Seeing Like a State: How Certain Schemes to Improve the Human Condition Have Failed*. New Haven, Connecticut: Yale University Press.

Sherjonov, M. 2003. Yangibozorning katta qo'llari. Newspaper article. *Yangibozor Ko'zgusi*, 15 August.

Snesarev, G. P. 1974. On Some Causes of the Persistence of Religio-customary Survivals among the Khorezm Uzbeks. In E. Dunn, and S. P. Dunn (eds.), *Introduction to Soviet Ethnography*, vol. 1, pp. 215–238. Berkeley: Highgate Road Social Science Research Station.

———. 1976. *Unter dem Himmel von Choresm, Reisen eines Ethnologen in Mittelasien*. Leipzig: VEB F. A. Brockhaus Verlag.

———. 2003. *Remnants of Pre-Islamic Beliefs and Rituals among the Khorezm Usbeks*. Berlin: Schletzer.

Songmoo, K. 1987. *Koreans in Soviet Central Asia*. Helsinki: Finnish Oriental Society.

Spoor, M. 1993. Transition to Market Economies in Former Soviet Central Asia: Dependency, Cotton, and Water. *European Journal of Development Research* 5 (2): 142–158.

———. 2003. Agrarian Reform in Post-Soviet States Revisited: Central Asia and Mongolia. In M. Spoor (ed.), *Transition, Institutions, and the Rural Sector*, pp. 47–60. Lanham, Maryland: Lexington Books.

——— (ed.). 2004. *Globalisation, Poverty and Conflict: A Critical 'Development' Reader*. Dordrecht: Kluwer Academic.

———. 2006. Uzbekistan's Agrarian Transition. In C. S. Babu, and S. Djalalov (eds.), *Policy Reforms and Agriculture Development in Central Asia*, pp. 181–204. Boston: Springer.

———, and O. Visser. 2001. The State of Agrarian Reform in FSU. *Europe-Asia Studies* 5 (6): 885–901.

Starr, S. F. 1999. Civil Society in Central Asia. In M. Holt Ruffin, and D. C. Waugh (eds.), *Civil Society in Central Asia*, pp. 27–33. Seattle: University of Washington Press.

Stevens, D. 2007. Political Society and Civil Society in Uzbekistan: Never the Twain Shall Meet? *Central Asian Survey* 26 (1): 49–64.

Subtelny, M. E. 1994. The Symbiosis of Turk and Tajik. In B. F. Manz (ed.), *Central Asia in Historical Perspective*, pp. 45–61. Boulder, Colorado: Westview Press.

Suyarkulova, M. 2008. Book reviews. *Central Asian Survey* 27 (3): 395–399.

Swain, N. 1992. *Hungary: The Rise and Fall of Feasible Socialism*. London: Verso.

———. 2003. Social Capital and Its Uses. *Archives Européennes de Sociologie* 44 (2): 185–212.

Tapper, R. 1990. Anthropologists, Historians, and Tribespeople on Tribe and State Formation in the Middle East. In P. S. Khoury, and J. Kostiner (eds.), *Tribes and State Formation in the Middle East*, pp. 48–73. Berkeley: University of California Press.

———. 1997. *Frontier Nomads of Iran: A Political and Social History of the Shahsevan*. Cambridge: Cambridge University Press.

Teichmann, C. 2007. Canals, Cotton, and the Limits of De-colonization in Soviet Uzbekistan, 1924–1941. *Central Asian Survey* 26 (4): 499–519.

Thelen, T. 2003. *Privatisierung und soziale Ungleichheit in der osteuropäischen Landwirtschaft: Zwei Fallstudien aus Ungarn und Rumänien*. Frankfurt a.M.: Campus.

Thompson, C. D., and J. Heathershaw. 2005. Introduction: Discourses of Danger in Central Asia. *Central Asian Survey* 24 (1): 1–4.

Thompson, E. P. 1991. *Customs in Common*. New York: New Press.

Thurman, J. M. 1999. *The 'Command-Administrative System' in Cotton Farming in Uzbekistan, 1920s to Present*. Indiana University Papers on Inner Asia 32. Bloomington: Indiana University.

Tingay, C. 2004. *Agrarian Transformation in Egypt: Conflict Dynamics and the Politics of Power from a Micro Perspective*. PhD dissertation, Freie Universität Berlin.

Tolstov, S. P. 2005 [1948]. *Following the Tracks of Ancient Khorezmian Civilization*. Tashkent: UNESCO.

Torsello, D. 2003. *Trust, Property and Social Change in a Southern Slovakian Village*. Münster: LIT Verlag.

Trevisani, T. 2007. Kolkhozes, Sovkhozes, and Shirkats of Yangibozor (1960–2002): Note on an Archival Investigation into Four Decades of Agricultural Development of a District in Khorezm. *Cahiers d'Asie Centrale* 15–16: 354–365.

Trushin, E. 1998. Uzbekistan: Problems of Development and Reform in the Agrarian Sector. In B. Rumer, and S. Zhukov (eds.), *Central Asia: The Challenges of Independence*, pp. 259–291. New York: M. E. Sharpe.

Tulepbayev, B. 1984. *Socialist Agrarian Reforms in Central Asia and Kazakhstan*. Moscow: Nauka.

Turaeva, R. 2008. The Cultural Baggage of Khorezmian Identity: Traditional Forms of Singing and Dancing in Khorezm and in Tashkent. *Central Asian Survey* 27 (2): 143–153.

Tursunov, S. N. 1996. *O'zbekiston qishloq xo'jaligini rivojlantirishining shartlari va pasayish omilllari (1946–1965)*. Tashkent: FAN Nashriyoti.

Veldwisch, G. J. 2008. *Cotton, Rice and Water: The Transformation of Agrarian Relations, Irrigation Technology, and Water Distribution in Khorezm, Uzbekistan*. PhD dissertation, Zentrum für Entwicklungsforschung, University of Bonn.

——, and M. Spoor. 2009. Contesting Rural Resources: Emerging 'Forms' of Agrarian Production in Uzbekistan. *Journal of Peasant Studies* 35 (3): 424–451.

Verdery, K. 1996. *What Was Socialism, and What Comes Next?* Princeton, New Jersey: Princeton University Press.

——. 1998. Property and Power in Transylvania's Decollectivization. In C. Hann (ed.), *Property Relations: Renewing the Anthropological Tradition*, pp. 160–180. Cambridge: Cambridge University Press.

——. 1999. Fuzzy Property: Rights, Power, and Identity in Transylvania's Decollectivization. In M. Burawoy, and K. Verdery (eds.), *Uncer-*

tain Transition: Etnographies of Change in the Postsocialist World, pp. 53–81. Lanham, Maryland: Rowman and Littlefield.

———. 2000. Ghosts on the Landscape: Restoring Private Landownership in Eastern Europe. *Focaal—European Journal of Anthropology* 36: 145–163.

———. 2002 [1991]. Theorizing Socialism: A Prologue to the 'Transition'. In J. Vincent (ed.), *The Anthropology of Politics: A Reader in Ethnography, Theory and Critique*, pp. 366-386. London: Blackwell.

———. 2003. *The Vanishing Hectare: Property and Value in Postsocialist Transylvania*. Ithaca, New York: Cornell University Press.

———. 2004. The Obligations of Ownership: Restoring Rights to Land in Postsocialist Transylvania. In K. Verdery, and C. Humphrey (eds.), *Property in Question: Value Transformation in the Global Economy*, pp. 139–159. Oxford: Berg.

Visser, O. 2008. *Crucial Connections: The Persistence of Large Farm Enterprises in Russia*. PhD dissertation, University of Nijmegen.

Vlek, P., C. Martius, and J. Lamers. 2001. *Economic and Ecological Restructuring of Khorezm Region*. Project proposal, ZEF/BMBF, Bonn.

Wall, C. 2007. Peasant Resistance in Khorezm? The Difficulty of Classifying Non-compliance in Rural Uzbekistan. In P. Sartori, and T. Trevisani (eds.), *Patterns of Transformation in and around Uzbekistan*, pp. 217–240. Reggio Emilia: Diabasis.

———. 2008. *Argorods of Western Uzbekistan: Knowledge Control and Agriculture in Khorezm*. Berlin: LIT Verlag.

———, and P. Mollinga (eds.). 2008. *Fieldwork in Difficult Environments: Methodology as a Boundary Work in Development Research*. Berlin: LIT Verlag.

———, and J. Overton. 2006. Unethical Ethics? Applying Research Ethics in Uzbekistan. *Development in Practice* 16 (1): 62–67.

Wang, F. 1994. The Political Economy of Authoritarian Clientelism in Taiwan. In L. Roniger, and A. Güneş-Ayata (eds.), *Democracy, Clientelism, and Civil Society*, pp. 181–206. Boulder, Colorado: L. Rienner.

Weber, M. 1972 [1921–1922]. *Wirtschaft und Gesellschaft*. Tübingen: J. C. B. Mohr (Paul Siebeck).

Wegerich, K. 2006. A Little Help from My Friend? Analysis of Network Links on the Meso Level in Uzbekistan. *Central Asian Survey* 25 (1–2): 115–128.

———. 2010. *Handing over the Sunset: External Factors Influencing the Establishment of Water User Associations in Uzbekistan*. Göttingen: Cuvillier Verlag.

Wegren, S. (ed.). 1998. *Land Reform in the Former Soviet Union and in Eastern Europe*. London: Routledge.

Wehrheim, P., and C. Martius. 2008. Farmers, Cotton, Water and Models: Introduction and Overview. In P. Wehrheim, A. Schoeller-Schletter, and C. Martius (eds.), *Continuity and Change: Land and Water Use Reforms in Rural Uzbekistan*, pp. 1–15. Halle: Leibniz-Institut für Agrarentwicklung in Mittel- und Osteuropa (IAMO).

Willerton, J. P. 1992. *Patronage Politics in the USSR*. Cambridge: Cambridge University Press.

Wittfogel, K. A. 1957. *Oriental Despotism*. New Haven, Connecticut: Yale University Press.

World Bank. 1999. *Consultations with the Poor: Uzbekistan*. Participatory Poverty Assessment in Uzbekistan for the World Development Report 2000/1, prepared by EXPERT Centre for Social Research (Uzbekistan).

Xolliev, A. 2003. Agrar Soxaning fan-texnika asosida rivojlanishi va iqtisodiy samaradorligi. *O'zbekistonda Ijtimoiy Fanlar* 4: 10–18.

Yaroshevski, D. B. 1994. The Central Government and Peripheral Opposition in Khiva, 1910–24. In R. Yaacov (ed.), *The USSR and the Muslim World: Issues in Domestic and Foreign Policy*, pp. 16–39. London: George Allen and Unwin.

Yuldoshev, M. Y. 1959. *Xiva Xonligida feodal yer egaligi va davlat tuzilishi*. Tashkent: O'zbekiston SSR davlat nashriyoti.

Yurkova, I. 2004. *Der Alltag der Transformation: Kleinunternehmerinnen in Usbekistan*. Bielefeld: transcript Verlag.

Zadykhina, K. L. 1952. Uzbeki del'ty Amu-Daryi. In S. P. Tolstov, and T. A. Zhdanko (eds.), *Trudy Khorezmskoy arkheologo-etnograficheskoi ekspeditsii*, Zadykhina, K. L. 1952319–436. Moscow: Izdatel'stvo Akademii Nauk SSR.

Zanca, R. 1999. *The Repeasantization of an Uzbek Kolkhoz: An Ethnographic Account of Postsocialism*. PhD dissertation, University of Illinois, Urbana-Champaign.

———. 2000. Intruder in Uzbekistan: Walking the Line between Community Needs and Anthropological Desiderata. In H. de Soto, and N. Dudwick (eds.), *Fieldwork Dilemmas: Anthropologists in Postsocialist States*, pp. 153–171. Madison: University of Wisconsin Press.

———. 2010. *Life in a Muslim Uzbek Village: Cotton Farming after Communism*. Boston: Wadsworth, Cengage Learning.

Zhao, B. 2009. Land Expropriation, Protest and Impunity in Rural China. *Focaal—European Journal of Anthropology* 54: 97–105.

Index

Abashin, S. 5, 35n, 42, 50n, 221, 222
absentee landlordism 179
accountability 130, 204
agency 3, 67, 183
agrarian; question 2, 5, 9, 222; transition 1, 6, 171, 195, 226
agricultural cycle 20, 21, 42, 122, 128-9, 211, 216
agriculture; dual structure of 99; female engagement in 205; in Yangibozor 22; industrial 65; newcomers to 179-80
agronomist 93, 172, 194; as large *fermer*s 204; chief 68, 89n, 115; interviews with 69, 143; *rais* with background as 83, 180
Allina-Pisano, J. 6, 27n, 113, 116, 123
Amudaryo 23, 31, 34, 80, 150; [as a pseudonym for a former collective farm] 170, 175, 177
anarchy 205
Andijon 8, 14, 214
arenda 69, 110
army 34, 72, 142
auction 103, 112-3, 119, 124, 159
authority; elderly as reference group of 40; 'fatherly' 177; in the family 44-5, 62; local 156, 190; notions of 62, 126; of the *brigadir* 203; of the *fermer*s 215; of the *hokim* 179; patriarchal 45, 50, 60; *rais* as symbol of 180
avlod 46, 139, 162
Azerbaijan 206

backwardness 38-9, 88
ball-bonitet 109-10, 159
bank 22, 27n, 63, 105, 121, 160; and district *hokim* 172; *fermer*s' bank accounts 109, 149, 185; hampering cash withdrawal 111, 149; loans for *fermer*s 186, 209; reform 122
bankruptcy; and *fermer*s 116, 123, 127, 159, 181; and MTP 106
basmachestvo or Basmaci movement 220
bayram 42, 214n
bazaar 23-5, 51, 159, 179, 201; 'bazarkom' 167; crop sale on 111, 154; for petrol 33n; prices compared to retail shops 107, 200; sale of cattle and 120, 182
bey 38, 49-51, 57, 71, 144, 178n; descent of a 117-8, 201; expropriation 71; house 54; 'reverting to the time of the beys' 177
'big man' *or* 'big men' 21, 183, 224
blackmail 123, 179
Bolshevik 31, 31n, 35, 35n, 51, 117
boshqarma [agricultural branch of the district administration] 22, 22n, 27n, 108, 150, 184, 194, *see also* district department (for agriculture)
Bourdieu, P. 204
Bo'ston 46, 54, 59, 60, 79, 79n, 86, 88, 110, 137-8, 142, 147, 162,
Bo'zqal'a 79, 79n, 126, 214-7
Bog'olon 34n, 58, 59, 79, 79n, 92, 126n, 203, 212
Bog'ot 28n, 32, 111, 141
Brezhnev, L. 11, 71, 75, 75n
brigade; and *elat*s 55; and kinship 164; and sharecropping 81, 181; average size in Khorezm 69, 69n; disbanding of 103-5, 135-9; family based 107, 116-7; in the *shirkat* 105, 130; in the Soviet period 61-2, 68-9, 71, 86-8, 203; prizes in the 87; role in the command system 124-5; working conditions in the 139, 146
Bukhara 31n, 34, 35, 38, 38n, 42, 210
bureaucracy; agricultural 10, 20, 82, 188, 190, 205, 224

business; collective land as a private 70; cotton 119, 168; criminal 185; land privatisation as a 114; plan of *fermer*s 102, 108, 114, 182, 194; public employees entering 25; rice as a risky 123
businessmen 29, 179, 180, 204
buva 45, 47, 50-1, 56-7, 162

cadastre 28, 59, 60, 102, 143n
capital; and cotton farms 152; and large farms 179, 204; and politics 3, 177, 183; and rice farms 151, 154; bureaucratic 204, 205, 207, 224; farmers deprived of 114; farmers' lack of 116; foreign 33; return to agriculture of 118; social 179, 204; stock 68, 101
capitalism 44, 195
capitalist relations 51, 71, 191, 223
cash; crop 25, 154, 161, 183; *fermer*s' lack of 154, 159-60; land transactions involving 110-2, 114, 161, 174; need 123, 141-2, 217; non-cash payment 25, 140, 151; payments in 110, 146, 151, 161; rural households' lack of 179, 199, 201; shortage in the district 24, 107, 182
cattle; as form of saving 140; before mechanisation 58; breeding 32, 86, 86n; brigade 69, 69n; farms 120; owned by *fermer*s 151, 182
China 4, 14, 216, 225n, 226
citizen 18, 22, 102, 102n, 108, 191
claim; family 118; justice 202; legal 171; to control of farm yields 178; to exclusive grazing rights 140; to land 116-7, 173, 211; to pre-collectivisation ownership 201
clan 151, 166-7, 176-7, 190, 222
class; antagonistic classes 49, 50; class-like hierarchy 137; divide 50; lower 118, 198, 200; new elite recruited from lower 118; of rural notables *or* elites 120, 153, 216, 225-6; of small *fermer*s 181; of 'working poor' 7; relationship in the making 142; society 2; upper 162
coercion; climate of 66; district authorities use of 98, 127-31; means of 170
collective farms; amalgamation in Yangibozor 72; as a permanent anomaly 66; members of 58, 67, 68, 71-2, 87, 103, 122, 163; peculiarities in Central Asia 67-69, 71
collective land 22, 99, 108, 110, 113, *see also* land, collectively owned; freely marketable crops on 110; illegal use of 108; sub-leasing of 70, 176
collectivisation; in Khorezm 57, 78; perception of 209; trauma of forced 44; Uzbek peasant before 198
colonial; broker 183, 183n; late Soviet Uzbekistan as a 'colonial' context 168
command; administrative-system 28, 66, 100n; economy 14; hierarchy 15, 17, 84, 98, 114, 123-5, 148, 171, 174, 209-11, 216, 223
commission; farm size optimisation 106n; land and farm establishment 108, 113, 137, 158; privatisation 113, 113n, 115-7; revision 58, 83
committee 89n, 90, 162; executive 69, 71, 73n, 90, *see also ispolnitelniy komitet*; *mahalla* 73; neighbourhood 27n, 40-1, 45, 73, *see also elat*; statistical 22n, 27n, 28n, 31n, 77
communism 42, 94-5; good communist leader 94, 221
Communist Party 57, 75n, 216; Central Committee of 65, 70n, 81; of the Soviet Union 70n, 79n; of Uzbekistan 55n

conflict; between *dehqons* and *fermers* 219; conflicting views over legitimate leadership 94; conflictual relationship with the authorities 155, 177, 186, 219; decollectivisation and 1, 8-9, 116, 219; difference between Soviet and post-Soviet dynamics of 188; in Yangibozor 170, 190; over cropping patterns 132; over rice 154; postsocialist 3-4

corruption 10, 75, 186-9; anti-corruption measures 189; 'honest' and 'dishonest' 95, 216; in the kolkhoz 74, 217; of the *rais* 202, 216, *see also* blackmail

cotton; advantages for producers 148-9, 154; and centre-periphery relations 95, 175, 221; and *plyonka* 159-60; and research 38; as source of revenue for the government 25, 99; barons 221; boll 144, 146, 148; burning of 90; buy undercover 182; Central Asia's role as cotton supplier 75; control group 129-30; export 99, 109; feast, *see bayram*; field 20, 47, 75n, 76, 99, 100, 149, 157, 161, 163; in Kyrgyzstan 206; oil 25, 143; pickers 48, 146, 148-9; plant 139, 144, 147, 194; prices for 101; procurement price for 106, 157-8; profit from 156; quality 109, 149; quota 183, 193, 215; seed 42, 159; sowing 42, 193; stalks 139, 143-5

cotton farm 67, 119, 151-2, 162, 182; and irrigation 150; labour relations on 143-6; typical problems of a 160

cotton harvest 24, 90, 109n, 140, 151, 146-150, 186; campaign 52n, 74, 109n, 148n

cotton ginnery; and *fermers* 121, 146, 149, 159, 175; cotton price per kilogram at the 109-10; privatisation 100n, 112; worker at the 140

cotton scandal; and data manipulation 11, 28, 90; and Soviet modernisation 220; and the cotton economy 188; 'victims' of 196, *see also* Uzbek affair

credit 8; and *fermers* 109, 148, 159-60, 209

crop, *see also* state crop; agreement with neighbour on 141; cash 25, 154, 161, 183; collection points 105; commanded 99, 107; decisions in the kolkhoz 69; *fermers'* decisions about 160; fodder 58, 110, 144; forage 25, 26; freely marketable 110-2, 123, 152, 180; rotation 85; strategic 99, 129, 155, *see also* cotton, wheat

cropping arrangements; and district authorities 123, 129, 193-4, 203, 206; confidential agreements on 173-4, 177; strategic considerations about 157

Dashog'uz 31, 34

debt; farmers inheriting 115; *fermer* families and 153, 157; of the *fermers* 122, 157, 185, 209, 213; of the *shirkats* 103, 115, 157

decision-making in agriculture 18, 48, 124, 129, 205n

defoliant 147, 149

deforestation 80

dehqon, *see also gektarchi*, peasant *pudratchi*; and subsistence 107, 120; and *tender* agreements 110; as residual category of reforms 216; attitude towards decollectivisation 198-202; families 106-7, 120, 142, 157, 199; household 112, 135-6, 157, 161, 199; household plots, *see tomorqa*; livelihood 138, 139-42; perception of betrayal among 209

district authorities; as policy implementers 195; control over *fermers*

130-1; control over input supply organisations 121, 177; *fermers'* shared interest with 149; problems with implementing decollectivisation 115, 118; relations between *dehqon*s, *fermer*s and 130, 166, 170, 189, 212, 215-6; tension between *fermer*s and 170, 177

district department (for agriculture) 22n, 27n, 121, 124, 159, 172-3, 178, 182, 184, 186, 193-4, 205, *see also boshqarma*

ecology 156-7; ecological degradation of the environment 99

economy; black-market *or* informal 67, 72, 74; command 14; cotton 65-6, 100n, 129, 188-9, 221; market 98, 191, 226; moral 1, 4-5, 11, 195; national 100; planned 26n, 183, 217; political 15, 95

education; decline of 8, 33, 201; difficult access to 117, 179; in the kolkhoz period 61, 71; privilege of the rich 201; religious 42; universal 47

educational background; as selection criteria for *fermer*s 113, 116; of large *fermer*s 204; of *rais* 180

efficiency 2; efforts to enhance 6, 69, 97; of agricultural production 210; of management 189

elat 40-1, 44, 52, 55, 56, 59, 62, 139, 147

elatkom 19, 27n, 41, 45, 73-4, 86, 115, 148n, 212

electricity 201; and irrigation 170; introduction of 54, 86; payment for 25, 108

elites; agricultural 66, 94, 187; legitimacy of the 168; loyalty of the 196; postsocialist 16; regional 167-9, 179, 196; relations with community 216, 220-2; social reproduction of 117-8

entrepreneur; agricultural 4, 14, 112, 225; political 177, 181-4, 224, *see also sponsor*

ethnic homogeneity 33-4

ethos; communitarian 221; market 163; traditional 5, 220

Europe 3n, 4, 5, 71

European 4, 7, 11, 14, 34, 67; Union 20

exploitation 2, 6; in the family 52; labour conditions and 133, 152; soil 85, 145

expropriation 6, 117, 209; during collectivisation 58, 71

factory 33, 87, 140, 200, 214

family; networks 27, 46, 133, 164, 206; traditional extended 54, 63

farm, *see also fermer*; *dehqonchilik* 119, 208; manager 18, 74, 129, 143, 176, 197, 204, 205n; member 151-2, 158, 160; naming of 56-7; orchard 72, 119-20, 136-7, 161-2, 210; private farming as burdensome business 160; profit 109, 123, 153, 156-7, 177, 181-3; rice 143, 156; supervisor 143, 151-2, 160, 182, *see also ish yurutuvchi*

farm consolidation policy 106, 106n, 108n, 120, 134n, 208, 225

fear; climate of 9, 18, 21, 131; of legal accusations 189; of losing land 207

Ferghana Valley 34, 36, 39, 42, 81n, 199

fermer; cotton growing- 120, 143-9, 150-1, 154, 158-60, 181; duties towards the nation 215; first-class and second-class 138; free-riding 156; household 133, 157; *katta* 142, 204, 215, 224; personally liable for performing plan 124; relationship with farm labourers 111, 139-40, 143-9, 151-3, 158-

60, 194, 201-2; rice growing- 155-6, 177; trouble for *fermer*s' business 213; *yangi* 179, 204, 224
Fermer and *Dehqon* Association *or* FDA 17, 20, 22, 22n, 105, 133, 146, 172, 172n, 181-9, 198, 206
fertiliser; bottlenecks in the availability of 101; 'land has become like a drug-addict' for 145; in the kolkhoz 68, 73, 216; rules for the use of 129; subsidised 109, 156-7, 176, 193
food 52, 98, 214
freedom 100, 110, 155, 187, 195, 206, 225; entrepreneurial 171, 219

gas 201; natural 8, 54, 86; no longer delivered 139
Gdlyan,T. and N. Ivanov 91-3, 196
Geertz, C. 7
gektarchi 143-7, 152, 163, 194, *see also dehqon, peasant, pudratchi*
Gellner, E. 221
gender 43-4, 47-8, 153, *see also* women
Georgia 171
Gorbachev, M. 69
groundwater 54, 109n, 141, 155, 157
Gurlan 23, 24, 32, 35, 78, 80, 90, 169, 184, 186

habitus 39, 94
hacienda 5, 129
hashar 52, 52n, 212
hokim; and steering group 130; city 169; regional 84, 169; role in land code 102-3; social origin of today's 117; spying for 183; uprising against 170; village 126-7; vision of the future 207-8
hospital 82, 146, 175, 212, 213
household; agreements between *shirkat* manager and heads of 103; average size of 46; debt and 115;

dehqon 112, 133, 135-6, 157, 161, 199; *fermer* 133, 157; income 67; rice and 111, 141-2; subsidiary plots 26, 48, 70, 99, 106-7, 134, 140, 155, 161, *see also tomorqa*; survey 19, 27, 46n, 199
Humphrey, C. 3, 33, 67, 72, 77, 179, 191, 197
Hungary 3-4, 71, 204

identity 8, 11, 177, 217; Khorezmian 36-38
ideology 4, 37, 75, 191, 195
ijara; contract *or* leasing contract 102-3, 106-8, 108n, 133, 155; in the khanate of Khiva 48; land 105
Ilkhamov, A. 8, 17, 37, 67-8, 99, 101, 127, 165n, 166-9, 171, 178-9, 181, 216, 224
'indigenous socialism' 95
inputs; agricultural 68, 101, 103, 109n, 122, 143n, 163, 172, 219; subsidised 109, 127, 154, 156, 176
Institute for Irrigation and Agriculture 158
international organisations 99
investment 14, 99n, 129, 168; by *fermer*s 179; by the state 81; district authorities' containment of 155; lack of 101; lack of alternatives for 154
irrigation; and cotton 67, 129, 150, 156, 160; and electric pumps 170; and rice 140, 150; channels 52n, 55, 61, 80; decisions about 160, 163; water wheels and 49
ish yurutuvchi 143, 160, *see also* farm supervisor
ispolnitelniy komitet 69

Jumaniyozov, M. 84-5
justice 75, 117, 195, 202, 223

Kandiyoti, D. 4n, 8-9, 12-3, 15-6, 24, 44, 48, 75, 99, 101, 153, 164, 167, 195-8, 217
Karakalpak Autonomous Republic *or* Karakalpakistan 31, 34, 80, 155-6
Kazakhstan 25, 111
Kehl-Bodrogi, K. 5, 35-7, 41, 56n, 118
KGB 118
khan 92-3, 116, 118, 178n, 190, 202
khanate of Khiva 24, 31, 35, 48, 51, 116n
Khiva 31-3, 35, 44, 55, 62, 117
Khorezm school 43, 51, 63
kinship 20, 41, 44-5, 50, 56, 118, 138-9, 142, 162-4, 187, 223-4; re-efflorescence of 164; 'territorial-kinship groups' 56, 143
kolkhoz; administration 69, 71, 177; as a 'total social institution' 72, 222; chairman 6, 36, 58, 62, 66, 68, 69, 73, 86, 167; difference between kolkhoz and sovkhoz 67-8; notables 71, 202; *rais* 6, 56, 58n, 62, 66, 68, 73-4, 83-4, 91-4, 110, 120, 124-5, 172, 179, 203; society 180, 197
kolxozchi 68, 136, 139, 153, 198, 199, 201, *see also* worker
Komsomol 40, 58n, 158
kontor 61, 81, 106n, 109n, 135, 138, 144, 208
Korean 34, 80, 80n
Krader, L. 44n, 46-8
Kyrgyzstan 137, 206

labour; child 9n, 148; farm labourer 111, 116, 129-30, 133, 143-6, 151-2, 160, 194, *see also* worker *or yo'llanma ishchi*; gendered division of 47-8; waged 25, 103, 198, *see also kolxozchi*
Lampland, M. 3, 66, 202, 204
land; appropriation by the kolkhoz notables 202; brigadier's 117; code 98, 102, 102n, 106, 172n, 198, 206; collectively owned 99, 108, 110, 113, 123, 137, 219; distribution 10, 17, 23, 29, 115, 134, 137, 162, 201-2; entitlements to 18, 25, 111, 123, 134, 137-8, 164; fallow 26, 61, 81, 139; *fermer*s returning their 211; 'hidden' 130; lease 102-3, 110, 113-4, *see ijara*; owned by the state 7, 97, 101-2, 209; register 114; rent 48, 107, *see pudrat*; restitution of 137; sub-lease 70, 103, 112, 176-7; tenure 7, 10, 48-9, 137; to buy land illegally 161
land measurer 29, 60n, 68, 106n, 114, 134, 144, 149, 158, 172, 179, 188, 211, *see also zemlemer*
land's 'masterlessness' 97
laqab 56, 56n
Leach, E. 224n
legal disputes in agriculture 123, 175, 182
legitimacy 4-6, 17, 41, 71, 190, 226; cultural 221; of personal networks 196; of the elites 168; of the MTP *rais* 126; state 216
Levin, T. 39-42, 84n

Madaniyat 56, 61, 68n, 72-3, 79, 79n, 83, 85-6, 110, 126-7, 143, 145-7, 202
mahalla 5n, 27n, 40-1, 43, 62, 73-4, 147, 200
Manguberdi, J. 37
marriage 34, 44-5
mechanisation of agriculture 44, 139
migration 61, 208, 212, 221; labour 24n, 48, 212; urban 67
militia 17, 19, 22, 131, 147, 149, 153, 173, 175-81 *passim*, 204, 212, 215, *see also militsiya*
militsiya 18, 22, 177
minifundia 107
Ministry of Agriculture 113

mistrust 18, 22, 152, 189
modernity 87-8, 220
monopoly 98, 109, 122, 122n, 124
Moscow 14-5, 38n, 75, 91-4, 168, 220
mosque 40-3
motor traktor parki 59-60, 63, 103-4, 134, 203, *see also* MTP
MTP; land 106, 119; *rais* 118, 124-7, 158, 172n, 174
Muslim 5, 12, 42, 67, 71, 220-1

nation; postsocialist 7, 165
national; delimitation 2, 31, 31n, 35; economic policy 6, 217; GDP 99; interest 129, 149, 189
nationalism 38, 221-2
Navro'z 36-7
network; kinship 41, 139, 162, 224; of input and marketing services 171; patronage 10, 11, 165-8, 176, 190, 196, *see also* patronage relation
nomenclatura 204
nostalgia 71, 201, 217

oqsoqol 41, 62
orchard 55, 72-3, 111, 136, 137, 142, 161-2, 210; auction of 112-3, 124; farm 72, 119-20, 136-7, 161-2, 210; land 112
Otajanov, A. 203
otalik 177-8
ownership 3-4, 18, 222, 225-6; exclusive 195; formal 177; land 7, 23, 48, 97-8, 101-2, 138; legal 171-2; mixed 112; of assets 116; private land 51, 97, 209, 225; rent 48; state 97, 101-2
Oyoqdo'rman 84, 126n, 135-7

palov 36, 88, 154, 203, 214
patronage relation 11, 74, 176, 181, 190, 224; business oriented patron-client relation 178
peasant; *see also dehqon, gektarchi, pudratchi*; access to firewood 139; resistance 9; 'retraditionalised' 223; subsistence needs 99; world 20, 50
pension 71, 73, 108, 158, 210
perestroika 69, 97, 196, 247
pesticides 85, 203
petrol 33n, 73,
pokaz 20, 42, 126n, 128, 211
Polanyi, K. 191
posbon 147
poultry 183
poverty 7, 33, 48, 49, 198
power relations 10, 49, 98, 177, 191, 220; centre-periphery narratives of 166; pre-reform 6, 127; structure of 1, 219; traditional 63
President 93, 102n, 106n, 126, 186, 206, 212
prestige 73, 91-2, 118; leadership's loss of 208; prestigious jobs 117; prestigious origin 50n
privatisation 3-4, 7, 14, 23, 25, 97, 111-4, 117, 122-4, 197, 202-3, 208; commission 113; of orchards 112-3; of tractors 112, 122
profit; and farm sponsoring 177; distribution of 151, 162; *fermer*s squeeze dehqon for 202; from rice and cotton compared 156, 175; from unofficial land use 111; illegal 173
property 2-6, 48, 51, 70n, 172, 204; collective 74; free-riding on state owned 88; fuzzy 171; state 88, 112
protest 3, 8-9, 117, 156, 170, 222
pudrat 69, 102, 105, 107-8, 118n, 123, 125, 133, 135, 143n, 151, 161
pudratchi 115, 125, 141, 143n, 161, 199-201, 207, *see also dehqon, gektarchi*, peasant

pump 80-1, 160, 170

Qalandardo'rman 34n, 59-61, 126n, 137n
Qipchaq 35
Qizilqum 31
Qoraqum 31
qul 56, 56n

rais; as capitalist manager 66; decline of 180; legitimacy of the MTP 126; *raisbuva* 63; *raislik* 84-5, 92
ram fight 36, 88
Rashidov, Sh. 55n, 75, 187
redistribution 74, 137, 180; of land 17, 29, 201-2, 209; of property 51; socialist model of 191
reform; conservativism 99, 167; implementation 8, 25, 57, 123, 128, 186, 204, 207, 210; intensification 7, 165; land 6-7, 10, 18, 20-1, 28, 46, 97-8, 104, 118, 163, 194, 203, 219; of civil administration 125-7; plans 208; postsocialist 4, 97, 194; winners and losers 4, 142, 197, 204, 222
're-patriarchalisation' 164, 223-4
resistance; passive form of 168; peasant 9; to change 44; to reform 7; to state control 166
responsibility 19, 111, 128, 155, 210; *fermer*s' 219; for the family 144; individual 97, 99; lack of clearly defined 97; paternalistic 73, 142, 196, 210; towards the community 175, 195, 210-1, 216
rice; farm 143, 150-2, 156; households and 141-2; profits 169, 193; prohibition 155-6
rights; cropping 25; grazing 140, 219; human 18n; individual over land 162; to establish a farm 113; to life-long and inheritable possession 102

ritual 40-2, 67, 83, 211, 215; 'rebellion rituals' 190
Romania 3-4, 171, 223
Roy, O. 12n, 55n, 180, 190, 195-8, 217
rural society; and decollectivisation 8, 215-7, 223; class-like hierarchy in 137; peasants' place in 206, 208
Russia 3, 5, 25, 129, 140, 200-3

salinity 23, 109n, 116, 156, 159, 181
Samarkand 34, 42, 86
Sart 35, 35n; Khiva Sarts 35
Sazonova, M. V. 34-5, 43, 45-7, 49-53, 55-6, 223
school 25, 72; and cotton 147-8, 214; and *fermer*s 212; director 148, 204; pupils 82, 87; teacher 115
Scott, J. 9, 66, 202
seeds 58, 68, 107, 109, 145, 170; and vegetable-growing sovkhoz 86; cotton 42, 159
sel'soviet 40-1, 59-61, 73-4, 125-6, 158
sharecropping 103, 108, 110-2; and *fermer*s 111, 151, 176, 179, 181; as a substitute form of remuneration 25, 145; in the kolkhoz 110
shareholder 6, 100n, 103, 111, 129, 157, 204
shirkat; 'de-shirkatisation' 123; land 108, 112-3, 118n, 125, 137, 140, 158, 186; *rais* 108, 114, 124-5, 158; *rais*' and *brigadir*'s role in 108
sho'ro 41, 74, 126-7, 138, 153, 158
Shovot 24, 32, 78, 110, 174
Snesarev, G. P. 43-8, 52, 55-7, 62, 143
social inequality 7, 8, 74, 197, 219
social order 219
soil 23, 85, 86, 109n, 145, 156, 159, 181, *see also* salinity
solidarity 22, 40, 75, 177, 197, 222-3; family 163; groups 166, 190, 217

Soviet; authorities 62, 220; ethnographers 13, 44, 52, 220; law 73; modernisation of the countryside 59, 62-5, 220
sponsor 176-8, 180-1, 190, 224
state; 'abandoned by the state' 206; and families 163; as an all-round caretaker 216; as monopolistic buyer 68, 99; care for cotton-growing *fermer*s 154; Uzbekistan as an agrarian 129
state crop; and families 107; and *fermer*s 104, 120, 206-7; and productivity 101, 140; extemption from 129, 183; trade with 109
status; decline of *rais'* 180; difference between brigadier and *kolxozchi* 136; economic 154, 204, 212; *fermer*s' status in rural society 215; in the family 44, 52; legal *or* juridical 106-7, 119, 124, 143n
stratification; among *fermer*s 128, 153; continuity in 211; in the kolkhoz 62, 71-2; rural society falling apart in new 223-4; tripartite social 216; within farms 160
surplus 24, 68, 70, 87, 179
Surxondaryo 214

tabelchi 72
Tajikistan 67
tanob 48-50
Tashkent 15, 21, 33-4, 37-9, 42, 78n, 84, 108n, 112, 158-9, 181, 200, 204
taxation 48-9, 51, 71, 106, 175; indirect 101, 106n; land 100-1; 'one time taxation' *or chakana nalog* 175; water 100-1
theft 88-9, 113, 196, 202, 217, 222; of cotton 149; of fertiliser 152
Thelen, T. 3-4, 164
Thompson, E. P. 4
Tolstov, S. 34, 43, 45

tomorqa; constraints on 142; *dehqon* households and 45, 106, 136, 199; households sharing harvest of 46; income from 140; land measurer and 134, 149; land needed to provide new 105, 211; pooling of 162, 198; prioritised over *fermer*s' land 211; procedure to get 106n; rice prohibition and 155; selling the use of 111; women and 48, 161
to'qay 80, 80n
To'rtko'l 34
to'y 36, 40-1, 90, 139, 142, 151, 193, 198, 214-5, *see also* marriage, wedding
tractor; collectively owned 124; driver 48, 84, 142, 144, 159; fuel 33n, 146, 159; leasing of 158; new 122; of the MTP 122, 176; privatisation of 112, 122; quasi-monopoly over 122; shortage of 122, 122n, 160
trade 23-5, 33, 109, 120; licenses for 109; terms of trade of farming 123, 130, 160, 205; with cotton 149
trustees 85, 87
tsarist period 35, 48, 220
Turkestan 35, 35n, 38, 71
Turkmenistan 31, 33-4, 38

UNESCO 16, 27, 33
university 83, 148, 201; degree 204; fees 201; students 148, 148n, 214; teacher 30, 42, 176, 178, 201, 204
Urganch 24, 29-30, 32-4, 36-8, 42, 45, 78n, 80, 83, 148, 159, 169, 170, 177, 184, 201, 204; *rayon* 78, 84, 151
Uzbek affair 11, 65, 91, 168, 196, *see also* cotton scandal

vegetables; as second crop 110;

brigade 69; farm 119-20; processing factory 86-7; seed and vegetable-growing sovkhoz 86
Verdery, K. 2-3, 71, 116, 118, 137, 164, 171-2, 189, 191, 205, 223
village; administration 40, 58n, 61, 73-4, 126-7, 138, 147, 200, 212; bobby 147, *see also posbon*; booklet 34n, 58n; council 24, 106n, 126n, 135n, 137n, 213; *hokim* 126-7; mullah 42; registry books 138, 147; wealthy 120
villagers 33, 46; 'sell their cows to satisfy the city's appetite' 179

waste; fight against 69, 74, 88, 99; in the kolkhoz 74
water, *see also* groundwater, irrigation; and *hashar* 52n; cost of irrigation 25, 156; reforms 50, 57; shortage 8, 23, 116, 139, 145, 176; supply for irrigation in Yangibozor 23; usage norms 129; waste 28n; waterways 80
water user association *or* WUA 101, 101n, 104-5, 121
Weber, M. 184
wedding 20, 22, 40, 51, 140, 142, 151
weeding 144, 150-2, 163, 194, 203
welfare 3, 68n, 71, 87, 195, 209; national 155; social 8, 200
wheat; as a state-ordered crop 107, 127; contract 159; delivered for donation 146, 175; harvesting machine 143; imports 98; on the *tomorqa* 141, 149; payment in 142, 159, 182; winter 32, 110, 120, 128, 140, 145, 148, 150, 213
Wittfogel, K. 12n
women 34, 43, 47-9, 153, 161, 194, 202; *fermer* 158, 181, 205, 205n; peasant 170
worker; brigade 69, 107, 203; farm 125, 129-30, 143, 146, 151-2, 160, 193-4; kolkhoz 7, 68, 70, 74, 81, 86n, 110, 118n, 139, 148, 196, *see also kolxozchi*; *mardikor* 152; seasonal 24, 160; sovkhoz 70; urban 81, 164
World War II 44, 71, 78n, 181

xalq 6, 84, 92, 95, 148, 200
Xalqobod 79, 79n, 116-7, 126n, 134-7, 181
xayriya 146, 212
Xazorasp 32, 89n
Xinjiang 14-5
Xonqa 32-3, 35, 80n, 84

Yangibozor city 42, 79, 140
yo'llanma ishchi 143-4, 146, 151-2, 160, *see also* worker
yoshulli 22, 62, 177, 193-4, 209-11

zemlemer 68, 114, 158, *see also* land measurer